HIGH RESOLUTION RADAR IMAGING

HIGH RESOLUTION RADAR IMAGING

Dean L. Mensa

Pacific Missile Test Center
Point Mugu, California

International Standard Book Number: 0-89006-109-2
Library of Congress Catalog Card Number: 81-71048

To my wife, Judith
and
to my mother, Ida

Table of Contents

Table of Contents

Table of Illustrations

Preface

A decade ago, microwave imaging with resolution of a few wavelengths required large investments for sophisticated, special purpose equipment. Today, owing to developments in microwave components, electronic instruments and computers, such measurements can be performed in laboratories instrumented with commercially available equipment. The book describes techniques for high-resolution microwave imaging of radar targets which can be implemented using such equipment. It is written for engineers with some knowledge of microwave instrumentations and Fourier transforms. The techniques described use object rotation to form circular synthetic apertures which can be steered and focused to obtain lateral resolution. When pulse compression and synthetic aperture processing are combined, two dimensional resolution can be obtained.

Theoretical limits of the imaging performance are derived and related to physical quantities and parameters of the signal processing algorithms. Analogies between doppler filtering, synthetic aperture formation and holography are presented to facilitate formulations in the contexts of Doppler processing, antenna theory, or coherent optics which may be more familiar to some readers. Mathematics are used to establish a theoretical basis but are kept to a minimum; wherever possible, physical interpretations of the processes involved are stressed. It is my hope that the examples presented will encourage other workers to apply these methods to image objects irradiated by any form of coherent wave propagation.

The content of the book is a result of research conducted in preparation of a doctoral dissertation at the University of California, Santa Barbara. The experimental results were obtained using facilities at the Pacific Missile Test Center, Point Mugu, California which were funded by the Targets and Range Systems Division of the Naval Air Systems Command. I am indebted to my colleagues, W. Yates and R. Schlapia of the Pacific Missile Test Center, for executing the computer programs and computer graphics and to my research advisor, Professor G. Heidbreder of the University of California at Santa Barbara, for many consultations, suggestions, and reviews.

Introduction

THE CONCEPT OF MICROWAVE IMAGING

Radar sensors respond to electromagnetic waves which are scattered when the wave propagation is disturbed by the presence of an object. The fields incident on the object induce a distribution of currents which in turn establish the scattered fields. For purposes of this work, a radar image is defined to be the spatial distribution of reflectivity corresponding to the object. Adhering to this definition, it follows that the radar image can be considered a collection of reflection coefficients assigned to an array partitioning the object space. This notion is consistent with the IEEE definition [1] which states that an image is "a spatial distribution of a physical property such as radiation, electric charge, conductivity or reflectivity, mapped from another distribution of either the same or another physical property."

In order to characterize the object properly, an image must provide a spatial, quantitative description of the object's physical property of interest with fidelity which equals or exceeds the discrimination limits of the eventual observing system. The complete characterization of complex objects may require several images which differ as a function of the viewing angle. Useful optical images present spatial distributions of optical reflectivity with sufficient detail for the object to be recognized

Figure 1-1. Examples of Terrain Imagery by Airborne Synthetic
Aperture Processing.

by a human observer. Similarly, a radar image presents a spatial distribution of microwave reflectivity sufficient to characterize the object illuminated. Although the similarity of microwave images to optical images can convey useful information, the quality of the microwave image should not always be judged by how closely it approximates the optical image, but rather by how faithfully it represents the spatial distribution of microwave reflectivity.

An essential feature of any high-quality image is resolution, the ability to represent distinctly two closely spaced elements of the object, the resolvable spot size being inversely related to the aperture size of the imaging system. Because the human visual system is equipped with apertures of the order of 10^4 wavelengths, an order of magnitude greater in terms of wavelength than the largest of microwave apertures, human observers are accustomed to perceiving images consisting of millions of resolvable elements; as a result, optical replicas of microwave images often appear primitive by standards of optical images. Nevertheless, microwave images convey unique information and, in some cases, can display object features not otherwise obtainable. Figure 1-1 is an example of microwave imagery obtained by synthetic aperture processing described in [2]. This type of image, generated by processing radar data collected from an airborne platform, provides a two-dimensional map of terrain features. Figure 1-2 is another example of microwave imagery obtained by raster scanning the spot of a focused antenna across the object and mapping the magnitude of the received signal for each spot position. The upper figure is a photograph of a scale model of a boat, the lower figure is the microwave image obtained using a wavelength of 3 mm.

Imaging methods can be divided into two major categories according to the process employed: in-place imaging, and object-motion imaging. In the first method, the image is derived from observations of the object held in a fixed attitude relative to the observer; the resulting image is derived from and can be uniquely associated with a particular object orientation. In the second method, the imaging process requires relative motion between object and imaging system; in some cases, the resulting image is derived from and associated with a range of object orientations. Both imaging methods are useful but proper interpretation of the images must take into account the method employed.

The objective of this book is the description of techniques for obtaining high-resolution microwave images, with particular emphasis on methods which minimize hardware requirements and exploit signal processing.

1 METER

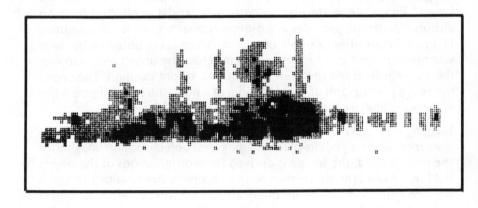

Figure 1-2. Microwave Image of Boat Scale Model by Focused Spot Scanning.

The principal applications for the microwave images are:

1. Analytical and physical simulation of radar target signatures for determining responses of radar sensors.

2. Diagnostic methods for the identification and alteration of radar reflectivity components of complex objects.

3. Nondestructive and noninvasive testing for imaging through media which support microwave propagation.

4. Object recognition systems which use the image as an identifier unique to a particular object.

The immediate application of the techniques to be considered is microwave imaging of reflective objects in a controlled environment. Typical applications employ microwave radiation with a wavelength of 3 cm to image objects with resolution of 2 to 3 wavelengths. The object dimensions are of the order of 100 wavelengths and the observation distance of the order of 1000 wavelengths. Although these are the intended applications, the imaging methods considered have general applicability to environments outside the laboratory and to other forms of coherent wave propagation. The direct application of these methods to acoustic imaging for medical diagnoses and non-destructive testing is feasible because modern instrumentations allow coherent measurements of acoustic fields.

The following paragraphs outline the major aspects of individual chapters and identify those parts which constitute original work.

Chapter 2 presents essential background material; it summarizes theories pertinent to coherent imaging which have been interpreted in terms of the research objectives. With the exception of some interpretations the contents of the chapter are not original and are traceable to major references.

Chapter 3 describes an experimental system and signal processing algorithms for obtaining two-dimensional (range, cross-range) images from experimental data. The system concept has been described in the literature; however, the practical implementation to achieve range resolution of less than two wavelengths constitutes a unique development. Original contributions in this chapter consist of analyses of the image degradations which result from Doppler processing. This process is analyzed in terms of an equivalent unfocused synthetic aperture which is subject to phase errors when steered through the object space. The analyses determine the limits of resolution, the maximum object size that can be imaged, and the resulting space-variant point-spread functions or spatial impulse responses. Experimentally derived images of complex objects are presented; among these are novel images of a human body. An additional unique aspect of the work is the extension to two-dimensional apertures synthesized by object rotation about two orthogonal axes. When combined with range sorting, this procedure allows three-dimensional resolution.

Chapter 4 develops the processing required to produce a focused synthetic aperture for CW irradiation. This allows the synthesis of a circular aperture which surrounds the object and results in a high degree of resolution. The process required to focus the aperture is also developed for the case of wide-band irradiation and point-spread functions for both CW and wide-band cases are derived. The effects caused by substituting several discrete frequencies for wide-band irradiating signals are determined. The feasibility of simulating wide-band imaging by multistatic measurements using CW irradiation is established and experimentally confirmed. The results of the focusing procedure applied to a wide-band irradiation are identical to those recently published in an application of optical processing of radar data. With this exception, the contents of this chapter are original. The achievement of high-resolution two-dimensional imaging using monochromatic irradiation and the analytic development of point-spread functions which have been experimentally verified constitute the significant contributions of this chapter.

In Chapter 5 the effectiveness of an iterative algorithm to extrapolate spatial spectra obtained from the focused synthetic aperture is tested. The algorithm is a two-dimensional adaptation of a one-dimensional version which has been described in the literature to extrapolate band-limited time functions. The strategy of the algorithm is, therefore, not original. The application to enhance two-dimensional images derived from synthetic aperture data, the algorithm implementation using two-dimensional FFT's, and the quantitative demonstrations of achievable results constitute the original parts of this chapter.

REFERENCES

1. Jay, F., *IEEE Standard Dictionary of Electrical and Electronics Terms*, New York, IEEE Inc., 1977, p. 361.
2. Jensen, H., L.C. Graham, L.J. Porcello, and E.N. Leith. "Side-Looking Airborne Radar," *Scientific American*, October 1977, pp. 84-95.

Fundamental
Imaging Techniques

In this chapter the basic techniques applicable to microwave imaging are developed. The treatment is limited to one-dimensional resolution in order to develop a conceptual foundation with minimum mathematical complications; extensions and combinations of the basic processes to achieve multi-dimensional imaging are considered in subsequent chapters. The emphasis in this chapter is placed on establishing fundamental concepts and relating results reported in the available literature to the meeting of the objectives.

Range Processing

The most basic method of radar imaging involves discrimination on the basis of range. The determination of range is accomplished by measuring the round-trip delay of the transmitted signal and computing distance using knowledge of the propagation velocity. The time delay between a distinct feature present in the transmitted waveform and recognizable in the received waveform is measured with electronic circuits. The range measurement is fundamentally a correlation process, the time delay which maximizes the correlation between transmitted and received waveforms corresponding to the two-way range. The most readily mechanized system for range measurement is the pulse radar which provides inherent time markers by the leading and trailing edges of the pulse. The round-trip propagation delay τ for a reflector at range R is 2R/c, where c is the propagation velocity. Thus a pulse of duration T corresponds to a range increment $\Delta R = cT/2$, and two points separated by greater distances will produce distinguishable pulses allowing resolution.

The resolution of closely spaced object features can be accomplished by narrowing the transmitted pulse width and increasing the system bandwidth B such that $BT \simeq 1$, thus yielding $\Delta R \simeq c/2B$. A time-bandwidth product approximating unity is inherent to the class of pulse radars in which a carrier is amplitude modulated by a pulsed waveform.*

The measurement of range can be treated more generally from a communication theory viewpoint. Two significant aspects of the measurement process are accuracy and resolution; the former constitutes the ability to provide an unbiased estimate of the absolute range, the latter is the ability to distinguish closely spaced objects. The standard deviation of a range measurement error in the presence of Gaussian noise is [1, 2]:

$$\sigma_R = c(8B^2 E/N)^{-1/2} \qquad\qquad (2\text{-}1)$$

where B is the signal bandwidth and E/N is the ratio of signal energy to two-sided noise power spectrum level.

Equation (2-1) indicates that an arbitrarily high degree of range accuracy can be obtained by using signals with large bandwidth or high energy. In imaging applications, however, the accurate determination of range is of limited significance, while resolution of range is of primary interest. If object points at different ranges are to be distinguishable at the receiver, the signal waveform must be as different from its shifted self as possible [3]. The autocorrelation function of the signal waveform s(t), expressed by:

$$R(\tau) = \int_{-\infty}^{\infty} s(t)s^*(t + \tau)\,dt \qquad\qquad (2\text{-}2)$$

must be as small as possible except in the vicinity of $\tau = 0$, the ideal being a delta function. Because the autocorrelation function and the power spectrum of a signal constitute a Fourier transform pair, a narrow

---

*Processing a coherent burst of repeated pulses constitutes an exception to the unity time-bandwidth product.

autocorrelation function corresponds to a wide signal bandwidth. The conclusion is that a high degree of range resolution always requires large signal bandwidths in accordance with:

$$\Delta R \cong \frac{c}{2B} \tag{2-3}$$

This is the generally accepted measure of resolution, the precise expression being dependent on the specific definition of resolution.

The use of matched filters and correlation receivers allows resolution commensurate with the signal bandwidth, independent of the particular waveform. The class of signals characterized by large time-bandwidth products has the advantage of providing reduction of the time-bandwidth product in the correlation process [4]. While for waveforms with BT $\simeq$ 1, the postdetection bandwidth is of the same order as the predetection bandwidth, waveforms with BT $\gg$ 1 allow a reduction in either bandwidth or time duration (or their combination) by a factor equal to the BT product. This feature facilitates the recording and postdetection processing of high-resolution signals and is important for applications where the image is formed by processing stored data.

Regardless of the method employed, range resolution allows the sorting of reflected signals on the basis of range. Figure 2-1 shows a radar observing a three-dimensional object; when range-gating or time-delay sorting is used to interrogate the entire range extent of the object space, a one-dimensional image of the object, termed a range profile, will result. The range sorted echo from each cell of a three-dimensional object consists of the vector sum of signals received from all object elements contained in the range cell, weighted by their respective amplitude and phase. The phase factor includes both the phase inherent to the reflection coefficient and the phase associated with the two-way propagation delay; the amplitude includes both the magnitude of the reflection coefficient and the spreading loss due to inverse square of the distance. The range profile is therefore a cell-by-cell sequence of the resultant magnitudes of the phasor sums of signals from all elements contained in the range-resolution cells.

Range resolution provides one-dimensional imaging restricted to a single dimension of the object. Without additional resolution in other dimensions it is relatively limited; it can, however, provide rudimentary imaging of special objects for which the range profile is of interest. The utility of this process will emerge when considered in combination with other methods of resolution. This process is a form of in-place imaging because the range profile can be obtained without relative motion between object and imaging system.

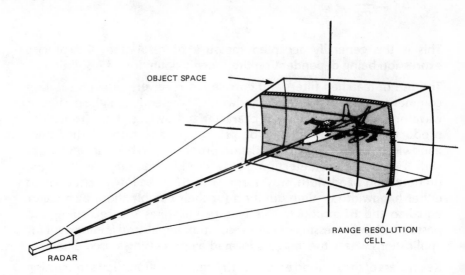

Figure 2-1. Range Mapping of Three-Dimensional Object by Range-Gated Radar.

Doppler Processing

Doppler processing is a widely used method for obtaining spatial resolution of objects irradiated with coherent waves [5-7]. The basis of Doppler processing is the observation that the frequency shift of signals reflected from a moving object is directly proportional to the radial component of velocity between the object and a stationary source and receiver. For the special case of a rotating reflector and a stationary source and receiver, the Doppler shift of the reflected signal is proportional to the lateral offset of the reflector, measured along an axis normal to the axis of rotation and to the line-of-sight. Doppler processing of signals reflected from a two-dimensional object rotating in a plane, therefore, yields resolution in cross-range.

The phase can also be expressed as a function of the rotation angle $\theta = \Omega t$ in which the angular rate of change of phase is:

$$f_\theta = \frac{1}{2\pi} \frac{d\phi}{d\theta} = \frac{-2}{\lambda} \frac{dr}{d\theta} = 2(d/\lambda)\cos\theta = 2x/\lambda \quad \frac{\text{cycles}}{\text{radian}} \qquad (2\text{-}7)$$

The frequencies in Equations (2-6) and (2-7), denoted by subscripts t and θ to indicate time and angle, respectively, are equivalent expressions; the angular and temporal frequencies are proportional to the instantaneous cross-range coordinate and differ only by the scale factor, Ω. Processing signals reflected from rotating objects recorded as a function of angle yields results identical to those obtained by processing signals recorded as a function of time for fixed rotation rate Ω. Both methods are termed Doppler processing in the literature; if a distinction is required, the former process may be termed temporal Doppler processing and the latter angular Doppler processing.

The equivalence of Equations (2-6) and (2-7) allows considerable flexibility in practical instrumentations which can be slanted toward temporal or angular processing. In the former method, the object is rotated at a constant angular rate. On-line processing is accomplished by a bank of contiguous frequency filters or by analyzing the sampled signal with Fourier transform processors of sufficient speed. Off-line processing can be performed by recording the signal for subsequent analysis; in either case, the signal is usually frequency-translated by ω_0 to retain only the complex envelope. In the latter method, the object is stepped in uniform angular increments and a series of static measurements is made of the phase and amplitude of the received signal. Providing that the Nyquist sampling criterion is met, a sequence of such measurements is identical to the phase history that would be observed with a continuous rotation.

The direct relation between Doppler frequency and cross-range allows spectral analysis of the signal to separate contributions of several reflecting points on the basis of cross-range. The frequency resolution for f_t varies inversely with the processing time duration T. Thus, $\Delta f_t \cong 1/T$ and, from Equation (2-6), the cross-range resolution is:

$$\Delta x = \frac{\lambda}{2\Omega} \Delta f_t \cong \frac{\lambda}{2\Omega} \frac{1}{T} = \frac{\lambda}{2\Delta\theta} \qquad (2\text{-}8)$$

where $\Delta\theta$ is the angular rotation during the processing time T.

Equation (2-8) indicates that fine cross-range resolution requires processing over a large interval of both time and angle. However, the instantaneous frequency of the echoes reflected from a particular point may vary considerably over a large processing interval, depending upon the angular position and rate of rotation of the object and the distance of the point from the center of rotation. The variation in the instantaneous frequency imposes limitations on rotational speed, maximum cross-range, and the number of pixels in the final image. To examine how the cross-range resolution is limited by the variation in instantaneous frequency, let the Doppler frequency space be divided into small uniform frequency intervals, Δf_t, corresponding to the cross-range resolutions cells of width Δx. Assume that the object can be enclosed in a circle of radius R. It is required that the movement of a particular reflecting point be confined to a single resolution cell during the processing interval T. If the rotational velocity, Ω is constant, the reflecting point will move most rapidly through a cross-range resolution cell when θ is $\pm \pi/2$. Thus, $\Delta x \geq \Omega RT$ or $\Delta x \geq R(\Delta \theta)$. Combining this with Equation (2-8) gives:

$$(\Delta x)^2 \geq \frac{1}{2} \lambda R \qquad\qquad\qquad (2\text{-}9)$$

If the restriction of Equation (2-9) is violated, Doppler frequency analysis will produce degraded imagery. To illustrate this degradation, we consider signals from various locations on a rotating object processed over a finite interval. Figure 2-3 shows the locus followed by each point object as it rotates over 0.15 radians; the observation distance, R_0, is 1000λ and the theoretical resolution according to Equation (2-8) id 3.3λ. The phase of the signal received from each object point considered individually is shown as a function of rotation angle in the left column of Figure 2.4; the amplitude is assumed constant and is not shown. The Fourier transform of the received signal for each case is shown in the right column plots of Figure 2-4. Because instantaneous frequency is equal to the rate of change of phase, the shape of the Doppler spectrum is determined by the phase history. In each case, the peak of the spectral response coincides with the cross-range coordinate of the point at the center of the processing interval; however, the non-linearity of the phase history causes spectral spreading with degradations which are more pronounced as the phase history deviates from linearity. As the phase nonlinearities increase beyond the values shown, the spectral responses no longer exhibit a single peak, thus producing severe image degradation.

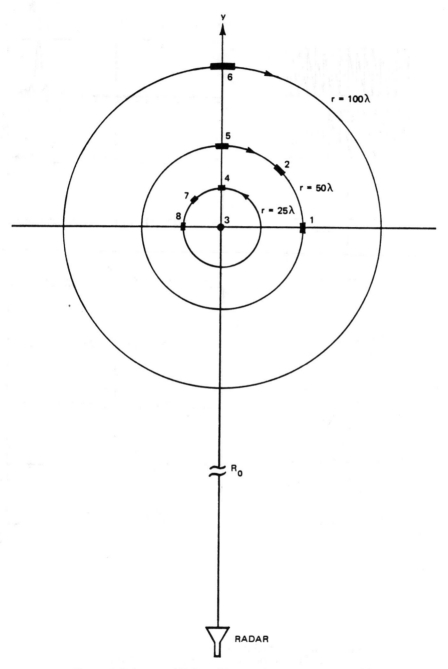

Figure 2-3. Locus of Points Observed on a Rotating Field.

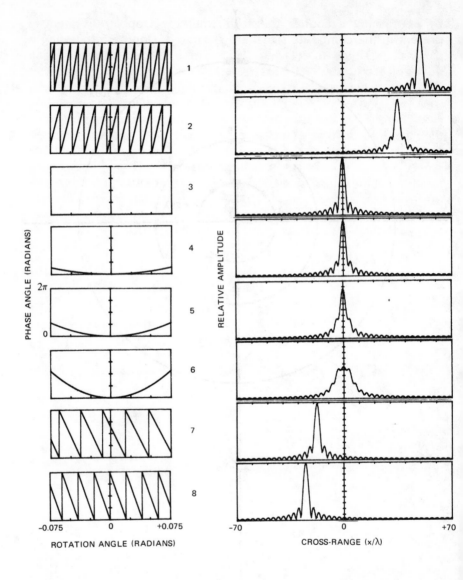

Figure 2-4. Phase of Signals Received From Point Objects
and Corresponding Cross-Range Spectra.

By performing a Fourier transform, multiple Doppler frequencies present in the composite signal are separated. Doppler processing is particularly suited to imaging rotating objects due to the relation between frequency and cross-range; however, other geometries involving relative radial motion among object elements can be utilized. Imaging a three-dimensional object with Doppler processing yields resolution along an axis normal to the axis of rotation and to the radar line-of-sight, as represented in Figure 2-5. Unlike the range-mapping case, in which range profiles could be associated with unique aspect angles, the cross-range profiles obtained by Doppler processing require that the object be rotated through a range of angles. Therefore, the cross-range profiles cannot be uniquely associated with a discrete aspect, but only with a range of angles necessary to induce the required relative motion, albeit small.

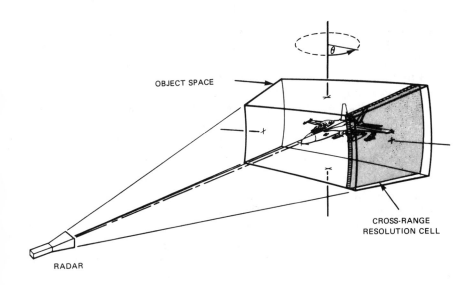

Figure 2-5. Cross-Range Resolution of Three-Dimensional Object by Doppler Processing.

Synthetic Aperture Processing

A high degree of resolution in the cross-range direction can be obtained by scanning a highly focused beam across the object. In order to achieve the minimum focused spot dimension, the aperture which forms the scanning beam must be focused at the object plane. The lateral extent of the focused spot is approximately:

$$\Delta = \lambda R_0 / D \tag{2-10}$$

where λ = wavelength

   D = aperture dimension

   R_0 = observation distance

If the focusing system is shift invariant, resolution of two adjacent object points lying in a plane normal to the line-of-sight can be expected if their separation is greater than the spot dimension. This is the Rayleigh resolution criterion [8].

For a fixed wavelength and observation distance, the resolution is improved by increasing the aperture dimension. A typical application of the imaging process requires the imaging of objects of the order of 3 m with resolution of the order of 6 cm from a distance of 25 m using a wavelength of 3 cm. The required aperture dimension given by Equation (2-10) is 25 m. Because the short measurement distance places the object in the near field, the aperture would require focusing in order to achieve the stated resolution. This requirement and the large dimension of the aperture constitute major disadvantages.

These difficulties can be avoided by synthesizing an equivalent aperture by sampling the fields present at discrete points on a surface conforming to that of the physical aperture. The fields are sampled with a small sensor which is sequentially stepped through incremental distances small enough to avoid aliasing. The subsequent coherent sum of the stored sampled values is equivalent to a signal that would be received by the physical aperture. By this method, an equivalent of a large physical aperture is synthesized by a sequence of field samples obtained with a small sensor. If the data from this aperture can then be processed to scan a focused beam on the target plane, an image of the target can be formed.

In the following section, the characteristics of synthetic apertures are established and related to those of physical apertures. The physical aperture with uniform illumination can be treated as an array of identical elements, excited simultaneously for transmission and summed coherently upon reception, as shown in Figure 2-6. The device connected to the array elements performs a summation on reception and equal power division on transmission. The response of the array to a unit-amplitude point object is the superposition of field values received by the individual elements; each receives a contribution from signals transmitted by all elements. The normalized complex field received from a distant point object by the nth element is given by:

$$v_n = \exp[-j2\pi\lambda^{-1}(d_n + d_1)] + \exp[-j2\pi\lambda^{-1}(d_n + d_2)] + \ldots \qquad (2\text{-}11)$$

$$+ \exp[-j2\pi\lambda^{-1}(d_n + d_N)]$$

$$= \exp(-j2\pi\lambda^{-1} d_n) \sum_{i=1}^{N} \exp(-j2\pi\lambda^{-1} d_i)$$

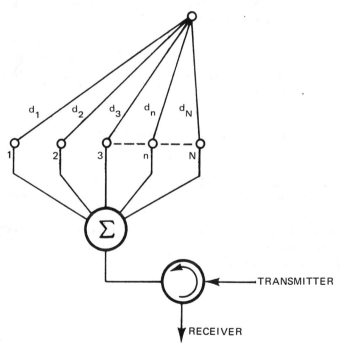

Figure 2-6. Physical Aperture Modeled as an Antenna Array.

The output of the entire array is the sum of signals received by each element.

$$v_0 = \sum_{n=1}^{N} v_n = \sum_{n=1}^{N} \exp\left(-j2\pi\lambda^{-1} d_n\right) \sum_{i=1}^{N} \exp\left(-j2\pi\lambda^{-1} d_i\right) \qquad (2\text{-}12)$$

$$v_0 = \left[\sum_{n=1}^{N} \exp\left(-j2\pi\lambda^{-1} d_n\right) \right]^2$$

The term inside the brackets is the one-way array response; it represents the response of the array if it were used to receive signals from a source located at the distant point. The response of the array when it is used for both transmitting and receiving (two-way) is the square of the one-way response. The cross-products generated in the squaring process represent the coupling terms associated with signals received by one element from transmissions by other elements. Equation (2-12) shows that the two-way response of the array is simply the square of the one-way response, an intuitively satisfying result when the array is viewed as a linear system obeying reciprocity.

The synthetic array is formed by sequentially transmitting and receiving with each individual element and subsequently coherently summing the received signals. The array geometry is identical to the preceding case. The response of the synthetic array will be:

$$v_0 = \sum_{n=1}^{N} \exp\left(-j4\pi\lambda^{-1} d_n\right) = \sum_{n=1}^{N} \left[\exp\left(-j2\pi\lambda^{-1} d_n\right)\right]^2 \qquad (2\text{-}13)$$

The response of equation (2-13) differs clearly from that of equation (2-12). The former represents the square of a sum, the latter represents the sum of squared terms. To illustrate the difference, the far-field responses of a real and synthetic array are determined. The array consists of N identical elements separated by d, and arranged in a line as shown in figure 2-7. The response of the real array is:

$$v(\theta) = \left\{ \sum_{n=-N/2}^{n=+N/2} \exp[j2\pi\lambda^{-1}nd(\sin\theta)] \right\}^2 \qquad (2\text{-}14)$$

Let $u = \sin\theta$, then

$$v(u) = \left[\sum_{n=-N/2}^{n=+N/2} \exp(j2\pi\lambda^{-1}ndu) \right]^2$$

$$v(u) = \left\{ \frac{\sin[\pi du\lambda^{-1}(N+1)]}{\sin[\pi du\lambda^{-1}]} \right\}^2$$

The response of the synthetic array is:

$$v(u) = \sum_{n=-N/2}^{n=+N/2} \exp(j4\pi\lambda^{-1}ndu) \qquad (2\text{-}15)$$

$$v(u) = \frac{\sin[2\pi du\lambda^{-1}(N+1)]}{\sin[2\pi du\lambda^{-1}]}$$

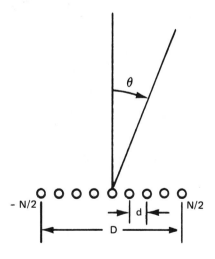

Figure 2-7. Linear Array of N Identical Elements.

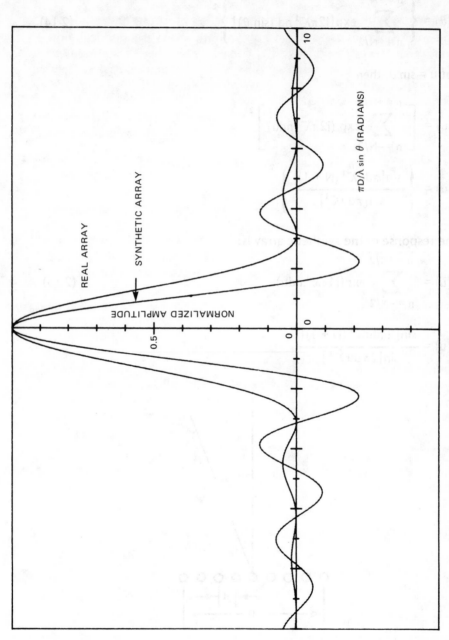

Figure 2-8. Two-Way Responses of Real and Synthetic Array.

The two responses are plotted in normalized form in figure 2-8. The one-way responses of both real and synthetic arrays are identical; however, the two-way responses differ considerably. This is a consequence of the coupling terms which arise when the array is used for both transmitting and receiving. The two-way response of the synthetic array is equivalent to the one-way response of a real array of twice the length. This feature, of significance in applications requiring maximum achievable resolution, is often overlooked when analogies between synthetic and real apertures are established.

Holographic Processing

One of the most significant developments in optical imaging has been the concept of holography. In this case, the imaging is performed by recording a pattern of light intensity which is used to reconstruct a replica of the light-waves associated with an object. A viewer observing these reconstructed waves perceives an image of the original object. The hologram encodes and stores a record of the waves diffracted from the object. Although microwave imaging concepts can be developed independently of holography, many of the intermediate steps in the recording and reconstruction processes for microwave imaging are directly analogous to holography. If recording microwave data and subsequently reconstructing an image from such data are performed optically, microwave imaging and holography become identical processes. From a review of the literature, it is apparent that the holographic process applied to coding and image formation from synthetic aperture microwave data was indeed the vehicle by which the two disciplines were joined to achieve significant contributions (9) to (11).

The concept of optical holography is briefly reviewed for the purpose of establishing fundamental similarities and differences between optical and microwave imaging. The intent is not to provide a comprehensive summary of optical holography, available in (12) to (14), but to consider the holographic process in the context of microwave imaging.

In optical holography, the object image information encoded on the hologram is the intensity pattern of the interference between the object and reference beams. The complete image information is contained in the complex amplitude (magnitude and phase) of the field associated with the object. Because phase sensitive optical storage media are not available, a recording of the interferogram of the object and reference wave is used to store the object complex field information in the form of a pattern of intensity. This elegant encoding scheme allows the subsequent retrieval (reconstruction) of the complex field. The generation of additional images and noise beams is a consequence of the intensity encoding process. The following development summarizes the procedure for recording and reconstructing a holographic image. If $U_T = U_O \exp[j\omega t + j\phi_0(x,y)] + U_R \exp[j\omega t + \phi_R(x,y)]$ is the superposition of complex object and reference waves incident on the hologram or x-y plane, the intensity of the light pattern on the plane is:

$$I = \overline{U}_T \overline{U}_T^* / 2 = \frac{1}{2} \ U_O \left\{ \exp[j\omega t + j\phi_0(x,y)] \right. \tag{2-16}$$

$$+ U_R \exp[j\omega t + j\phi_R(x,y)] \bigg\}$$

$$\cdot \left\{ U_O \exp[-j\omega t - j\phi_0(x,y)] \right.$$

$$+ U_R \exp[-j\omega t - j\phi_R(x,y)] \bigg\}$$

$$I = U_O^2 / 2 + U_R^2 / 2 + U_R U_O \cos[\phi_0(x,y) - \phi_R(x,y)]$$

The reconstructed wave, obtained by passing the reference wave through the recorded hologram, is expressed by:

$$\overline{U}_{REC} = I \ U_R \exp[j\omega t + j\phi_R(x,y)] \tag{2-17}$$

$$= (U_R/2) \exp[j\omega t + j\phi_R(x,y)] \ [U_O^2 + U_R^2]$$

$$+ (U_R^2 U_O/2) \exp[j\omega t + 2j\phi_R(x,y) - j\phi_0(x,y)]$$

$$+ (U_R^2 U_O/2) \exp[j\omega t + j\phi_0(x,y)]$$

The first two terms constitute noise beams associated with undiffracted light from the reference and object waves, the third terms represents a real image of the object, and the fourth term is a replica of the original object wave termed the virtual image.

The information contained in the holographic record can be obtained by electronic or computer processing of measured values of the object wave field by the system shown in figure 2-9. The real signals representing the object and reference waves are, respectively: $U_O \cos[\omega t + \phi_O(x,y)]$ and $U_R \cos[\omega t + \phi_R(x,y)]$

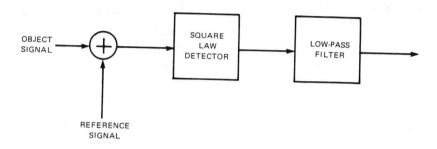

Figure 2-9. Signal Processing Model for Simulation of Optical Hologram Recording.

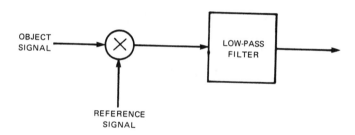

Figure 2-10. Alternate Signal Processing Model for Recording Image Information.

The output of the square law detector is:

$$V(t,x,y) = \left\{ U_O \cos[\omega t + \phi_O(x,y)] + U_R \cos[\omega t + \phi_R(x,y)] \right\}^2 \qquad (2\text{-}18)$$

$$= U_O^2 \cos^2[\omega t + \phi_O(x,y)] + U_R^2 \cos^2[\omega t + \phi_R(x,y)]$$

$$+ 2 U_O U_R \cos[\omega t + \phi_O(x,y)] \cos[\omega t + \phi_R(x,y)]$$

Expanding the above expression and removing the high frequency terms yields the output of the low-pass filter.

$$W(t,x,y) = \frac{1}{2} U_O^2 + \frac{1}{2} U_R^2 + U_R U_O \cos [\phi_O(x,y) - \phi_R(x,y)] \qquad (2\text{-}19)$$

The right side of equation (2-19) is identical to that of equation (2-16), indicating that the signal-processing model of figure 2-9 is equivalent to the process used in optical holography. The squaring and low-pass temporal filtering of real signals yields identical results to the intensity of the summation of complex signal inputs. Image reconstruction using a hologram formed with the recording process of figure 2-9 will yield the twin images and noise beams precisely as in the case of optical holograms.

A second method involving a multiplicative detection scheme is shown in figure 2-10, and analyzed as follows.
The output of the multiplier is :

$$V(t,x,y) = U_O \cos [\omega t + \phi_O(x,y)] U_R \cos [\omega t + \phi_R(x,y)] \qquad (2\text{-}20)$$

$$= \frac{1}{2} U_O U_R \cos [2\omega t + \phi_O(x,y) + \phi_R(x,y)]$$

$$+ \frac{1}{2} U_O U_R \cos [\phi_O(x,y) - \phi_R(x,y)]$$

Low-pass filtering suppresses the first term to the right of the equality leaving:

$$W(x,y) = \frac{1}{2} U_O U_R \cos [\phi_O(x,y) - \phi_R(x,y)] \qquad (2\text{-}21)$$

Equation (2-21) is similar to equations (2-16) and (2-19) except for the omission of the bias terms. Reconstruction of the image by multiplication with the reference yields:

$$U_{REC} = U_R \cos [\omega t + \phi_R (x,y)] \; \frac{1}{2} \; U_0 U_R \cos [\phi_0 (x,y) - \phi_R (x,y)]$$

$$= \frac{1}{4} \; U_0 U_R^2 \cos [\omega t + \phi_0 (x,y)] + \frac{1}{4} \; U_0 U_R^2 \cos [\omega t + 2\phi_R (x,y)$$

$$- \phi_0 (x,y)] \tag{2-22}$$

The first term is a virtual image constituting a replica of the object wave, the second represents a displaced real image of the object. The signal processing method of figure 2-10 is effective in removing the noise beams analogous to the undiffracted light in the optical hologram, but retains the twin images inherent to the holographic process.

The holographic imaging process is summarized in figure 2-11. Figure 2-11a shows the situation for conventional imaging. The object wave consists of light scattered by the object which propagates to the aperture of the imaging system and produces an image of the object. The disposition of the illumination source is arbitrary. The formation of the hologram, depicted in figure 2-11b, is a record of intensity of the interferogram of the object and reference waves. The object waves are identical to those in figure 2-11a, and the disposition of the reference waves is again arbitrary. The reconstruction of the hologram is shown in figure 2-11c. If the reference waves are precisely as in figure 2-11a, the reconstructed waves at the right of the hologram are identical to those of the original object in figure 2-11a. As previously shown, the waves to the right of the hologram contain additional images which can be separated from that of the object.

In microwave holography (15), the field amplitude scattered from an object coherently illuminated by a transmitter is mapped over a prescribed recording aperture by a coherent detector which is scanned over the aperture. The detected bipolar signal, representing the complex envelope of the time-varying field, is added to a bias level sufficient to make the resultant always positive. The resulting signal is used to produce a film transparency with an amplitude transmittance function which is real and positive. The area probed by the detector represents the hologram aperture, the reference signal for the coherent detector represents the reference beam, and the signal scattered from the object is the object beam.

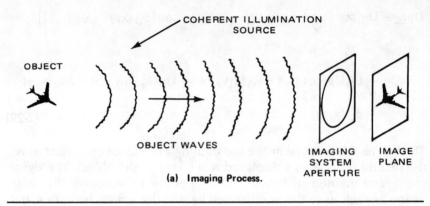

(a) Imaging Process.

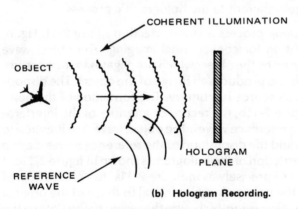

(b) Hologram Recording.

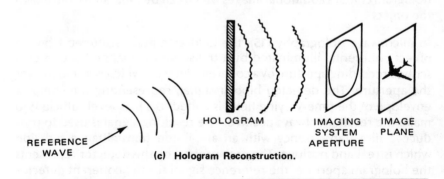

(c) Hologram Reconstruction.

Figure 2-11. Coherent Optical Imaging and Holography.

Microwave holography facilitates the independent scanning of the transmitter and receiver. This process, termed scanned holography, exhibits some advantages in resolution (16). When radar is used for imaging, the natural scanning process is one in which a collocated transmitter and receiver are made to scan a prescribed area. In radar terminology, this is the monostatic condition, in contrast with a bistatic condition in which the transmitter and receiver are moved relative to one another. The significant advantage of monostatic scanning is a two-fold improvement in resolution over the bistatic scanning. Figure 2-12 shows the bistatic and monostatic scanning cases. In figure 2-12a, the transmitter is stationary and the receiver is moved from position A to B. Signals received from two extreme locations on the scan differ in two-way path length by Δx; in figure 2-12b, the transmitter and receiver are both moved from position A to B and signals received from these two points vary in path length by $2\Delta x$. Because resolution is directly related to observed phase differences, the resolution performance is doubled for the monostatic scanned condition. As pointed out in (16), this type of scanning is equivalent to halving the wavelength used for bistatic scanning.

In any holographic imaging method, either conventional or scanned, the process is one of recording the diffraction pattern of the object falling onto a plane. The diffraction pattern is essentially the Fourier transform of the object illumination function; reconstruction of the image provides resolution in the cross-range direction similar to what would be achieved by an entrance pupil equal in size to the hologram. (This assumes a recording medium which is not limited in spatial frequency below the maximum spatial frequency of the diffraction pattern at the hologram.) Although the holographically reconstructed image is focused at all ranges, the image is basically two-dimensional; resolution in the axial dimensional being performed subjectively by human observers. This feature constitutes a significant difference between optical and microwave imaging; while two-dimensional imaging is satisfactory for visual observation of optical images, meaningful microwave images are generally required to be three-dimensional because radar sensors respond to relative spatial attitudes of object points in all three dimensions.

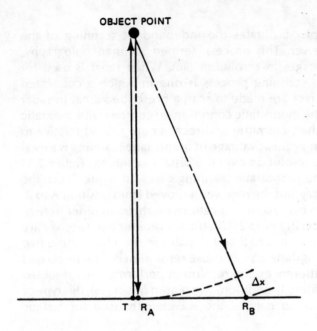

Figure 2-12a. Bistatic Field Scanning Geometry.

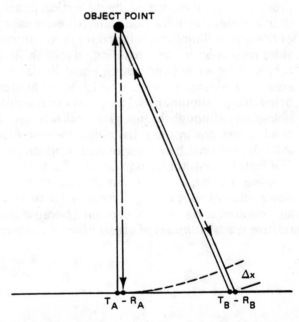

Figure 2-12b. Monostatic Field Scanning Geometry.

Relation of Imaging Methods

Imaging with microwave energy allows the achievement of direct resolution in the radial direction by using wide signal bandwidths and time delay sorting. This feature is completely independent of resolution in the cross-range dimensions which make use of diffraction phenomena.

Cross-range imaging by Doppler, synthetic aperture, and holographic processing lead to fundamentally similar results because each of these processes depends on the physical properties of fields diffracted from irradiated objects (17). The various approaches differ only in the initial viewpoint adopted to formulate the problem. Each of these processes utilizes the complex field incident on an observation surface which results from fields diffracted by the object. The angular interval, $\Delta\theta$, used to observe signals received from a rotating object for Doppler processing can be considered to form a synthetic aperture which is a circular arc subtending an angle equal to $\Delta\theta$. If the fields observed on this aperture were encoded on a recording medium, the stored information would constitute a hologram. The hologram could be used to reconstruct an image of the object which would be identical to that obtained by Doppler processing or by the response of the synthetic aperture focused and steered to all points of the object space. In the chapters to follow, we adopt the viewpoint which leads most directly to the required analysis.

REFERENCES

1. Skolnik, M. I. *Radar Handbook*, New York: McGraw-Hill Co., 1970, pp. 4-6.
2. Woodward, P. M. *Probability and Information Theory with Applications to Radar*, Oxford: Pergamon Press, 1964, p. 105.
3. Ibid., pp. 115-118.
4. Rihaczek, A. W. *Principles of High-Resolution Radar*, New York: McGraw-Hill Co., 1969, pp. 53-54.
5. Brown, W. M., and L. J. Porcello. "An Introduction to Synthetic Aperture Radar," IEEE Spectrum. p. 52, September 1969.
6. Brown, W.M., and R. Fredricks. "Range-Doppler Imaging with Motion Through Resolution Cells," IEEE Trans. Aerospace and Electronic Systems. AES-5, pp. 98-102, January 1969.
7. Hagfors, R., and D. Campbell. "Mapping of Planetary Surfaces by Radar," Proc. IEEE. Vol. 61, No. 9, pp. 1219-1225, September 1973.
8. Jenkins, F. A., and H. E. White. *Fundamentals of Optics*, New York: McGraw-Hill Co., 1950, p. 293.
9. Sherwin, C. W., J. P. Ruina, and R. D. Rawcliffe. "Some Early Developments in Synthetic Aperture Radar Systems," IRE Trans. Military Electronics, pp. 111-115, April 1960.
10. Cutrona, L.J., E.N. Leith, C.J. Palermo, and L.J. Porcello. "Optical Data Processing and Filtering Systems," IRE Trans. Information Theory, Vol. IT-6, pp. 386-400, June 1960.
11. Cutrona, L. J., E. N. Leith, L. J. Porcello, and E. W. Vivian. "On the Application of Coherent Optical Processing Techniques to Synthetic-Aperture Radar," Proc. IEEE, Vol. 54, pp. 1026-1032, July 1966.
12. Goodman, J. W. *Introduction to Fourier Optics*, New York: McGraw Hill Co., 1969.
13. Smith, H. M. *Principles of Holography*, New York: Wiley and Sons, Inc., 1969.
14. Goodman, J. W. "An Introduction to the Principles and Applications of Holography," Proc. IEEE, Vol. 59, No. 9, pp. 1291-1303, September 1971.
15. Farhat, N. H. "High Resolution Microwave Holography and the Imaging of Remote Objects," Optical Engineering, Vol. 14, No. 5, pp. 499-505, September 1975.
16. Hildebrand, B. P. and K. A. Haines. "Holography by Scanning," Journal of the Optical Society of America, Vol. 59, No. 1, pp. 1-6, January 1969.
17. Leith, E. N. "Quasi-Holographic Techniques in the Microwave Region," Proc. IEEE, Vol. 59, No. 9, pp. 1305-1318, September 1971.

Two-Dimensional Range, Cross-Range Imaging of Rotating Objects

We now outline a method which combines two processes described in the preceding chapter to produce two-dimensional microwave images. The object is rotated about an axis normal to the line-of-sight and combined time-delay sorting and Doppler processing are used to resolve the object space in two-dimensional resolution cells as shown in figure 3-1.

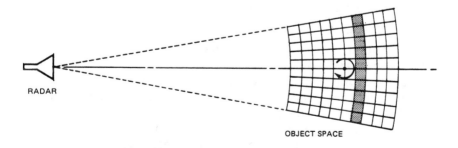

RADAR

OBJECT SPACE

Figure 3-1. Two-Dimensional Resolution of Rotating Object.

Resolution in range does not require object motion and is obtained by use of wide-band signals and time-delay sorting; by this method, back-scattered signals associated with iso-range cells, one of which is represented by the shaded area in figure 3-1, can be isolated. As the object is rotated, the signal associated with each range cell varies. For each range cell, the variations resulting from a finite rotation angle can be processed as described in the preceding chapter to obtain resolution in cross-range. By this method resolution in range and cross-range are independently obtained, the former by use of bandwidth and the latter by Doppler processing. The result of the combined processes is a two-dimensional image of the object formed in the radial (range) and normal to the radial (cross-range) directions lying in the plane normal to the axis of rotation.

A specific method for obtaining two-dimensional resolution is outlined in the following sections. Examples of images obtained by processing signals from an experimental microwave system are presented. Finally, the advantages and limitations of the imaging process and interpretations of the images are discussed.

Range Resolution

The required high degree of range resolution is achieved using large signal bandwidth and is given by the relation:

$$\Delta R = \frac{c}{2B} \qquad (3\text{-}1)$$

where ΔR = extent of range resolution cell

c = propagation velocity

B = signal bandwidth

As stated in the preceding chapter, the achievable resolution depends on the signal bandwidth and is independent of the particular waveform. Because the application being considered requires that received signals be coherently recorded for subsequent processing, a waveform with large time-bandwidth product is utilized. This feature allows the implementation of a correlation receiver which provides bandwidth compression, thus facilitating the instrumentation necessary to sample and store the received signals (1), (2). Perhaps the simplest correlation receiver of large time-bandwidth signals is implemented by using a linear-FM or chirp waveform. Figure 3-2 shows a simplified block diagram of an experimental system which operates by illuminating the entire object with a broad beam radiated from a small aperture antenna and receiving backscattered signals with a similar adjacent antenna. This instantaneous frequency of an RF carrier is linearly deviated about f_0 by $\pm B/2$ during a period T. The signals transmitted and received from a point object at range R are expressed, respectively, by:

$$e_T(t) = E_T \Pi \left(\frac{t}{T} \right) \sin(2\pi f_0 t + \pi B t^2 / T) \tag{3-2}$$

$$e_R(t) = E_R \Pi \left(\frac{t - \tau}{T} \right) \sin[2\pi f_0(t - \tau) + \pi B T^{-1}(t - \tau)^2] \tag{3-3}$$

The output of the mixer is described by the low frequency terms of the product of transmitted and received signals which, when using the relation $\tau = 2R/c$, reduces to:

$$e_0(t) = E_0 \Pi \left[\frac{t - R/c}{T - 2R/c} \right] \cos \left[\left(\frac{4\pi BR}{cT} \right) t + \left(2\pi f_0 - \frac{2\pi BR}{cT} \right) \frac{2R}{c} \right]$$

$$\tag{3-4}$$

The first term in the argument of the cosine function represents a frequency proportional to range, and the second term represents a range-dependent phase factor. For parameters of the intended application, T>>2R/c; that is, the signal duration greatly exceeds the signal round-trip travel time, therefore, the last phase term in equation (3-4) is negligible and the term in the first brackets can be well approximated by t/T. Equation (3-4) can thus be properly approximated by:

$$e_0(t) = E_0 \Pi(t/T) \cos [(4\pi BR T^{-1} c^{-1}) t + (4\pi f_0 c^{-1}) R] \tag{3-5}$$

Because the signal frequency is proportional to range, Fourier analysis of the signal in equation (3-5) provides a measure of range. If the object consists of several reflecting points, the composite signal will be a linear combination of signals of the form of equation (3-5) and, by the superposition property, the Fourier transform will accomplish range sorting. By this process, the object space is subdivided into range resolution cells.

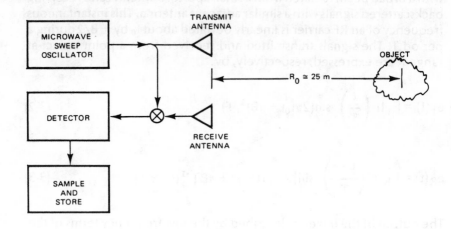

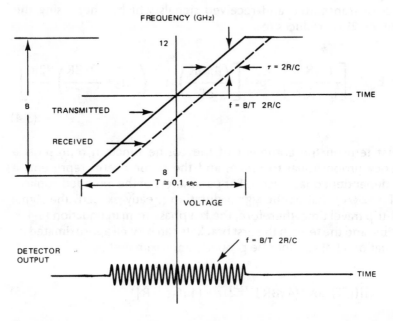

Figure 3-2. Linear-FM Microwave System Instrumentation.

The second term in the cosine argument of equation (3-5) is a phase angle proportional to R and to the frequency of the undeviated carrier. This phase term, essential for subsequent cross-range processing, is identical to that which would be observed with a CW measurement using frequency f_0. The range sorting process, therefore, preserves the phase information which is to be employed for cross-range processing.

The magnitude of the Fourier-transformed signal provides a range profile of the object with a resolution which is determined by the signal duration, T. The temporal frequency associated with an object at range R is:

$$f = \frac{2BR}{cT} \tag{3-6}$$

The spectral resolution, Δf, of a sinusoid observed over a period T is of the order of:

$$\Delta f = \frac{1}{T} \tag{3-7}$$

Combining equation (3-6) with (3-7) and associating the spectral resolution Δf with the range resolution ΔR, yields the relation invoked previously:

$$\Delta R = \frac{c}{2B} \tag{3-8}$$

Signal processing for range resolution consists of sampling the output of the mixer and performing spectral analyses using discrete Fourier transform (DFT) algorithms. Details of the computations are deferred to a subsequent section. The theoretical impulse response (point spread function) of the system is shown in figure 3-3. Reduction of the range sidelobes can be accomplished at the cost of slightly degraded resolution by applying conventional window functions to the time waveform (3). Figure 3-4 shows the theoretical impulse response resulting from weighting the signal waveform with a Hanning window.

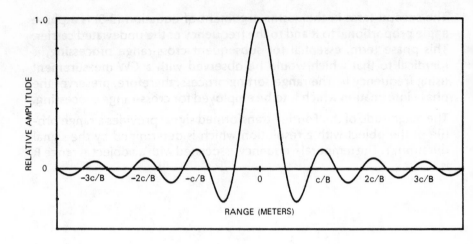

Figure 3-3. Range Response of Point Reflector (Rectangular Window).

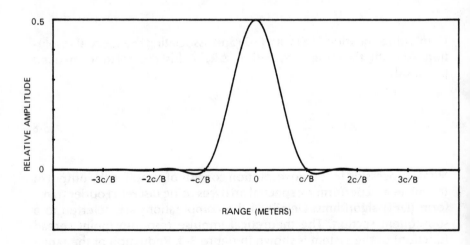

Figure 3-4. Range Response of Point Reflector (Hanning Window).

An experimental system operating at a center frequency of 10 GHz (λ = .03) with a bandwidth of 4 GHz was assembled and tested to obtain high range-resolution images. The period of the frequency sweep is approximately 100 ms resulting in a mixer output signal frequency of 8 KHz for a measurement range of 30 m. When the 4 GHz bandwidth signal, backscattered from a point object, is received, mixed, and Fourier transformed, the result is a spectral response near 8 KHz, with a bandwidth of approximately 10 Hz; this is a bandwidth reduction factor of 4×10^8, as given by the BT product. The time-bandwidth product of 4×10^8 is a relatively high value by practical standards. In order to achieve such high bandwidth compression ratios in this type of receiver, it is necessary that the frequency sweep by highly linear. The experimental system employs a subsystem for maintaining sweep linearity. This feature is not a significant aspect of the research and is reported in (4).

Experimentally obtained range responses to a point object are shown in figures 3-5 and 3-6. Figure 3-5 shows the magnitude of the spectral lines computed by the Fourier transform, and figure 3-6 is the envelope of the spectral lines. The data were processed using a Hanning window; some of the sidelobe structure is due to residual nonlinearities in the frequency sweep. Resolution is demonstrated by the response to a dual-point object in figure 3-7. Figure 3-8 shows a collection of range profiles taken for 1-degree increments of aspect angle of a corner reflector; the range profiles display the angular response of the reflector and confirm the impulse nature of the range response. Figure 3-9 shows an individual range profile for a relatively complex shape of a missile body. Several scattering centers can be associated with the physical structure. Figure 3-10 shows an ensemble of range profiles for the missile body at a number of aspect angles.

The range profile of the object obtained with the linear-FM system constitutes a one-dimensional image of the system; the magnitude of the range profile for a given range is the result of the coherent summation of returns from all object elements contained in a particular range cell.

Cross-Range Resolution

As indicated in the preceding chapter, a high degree of resolution in cross-range can be achieved by forming a large synthetic aperture.

The aperture is synthesized by the process of scanning the sensor over a surface, storing the received signals, and subsequently coherently processing these signals. We assume a *priori* that resolution in range has been accomplished by time-delay sorting methods and we therefore

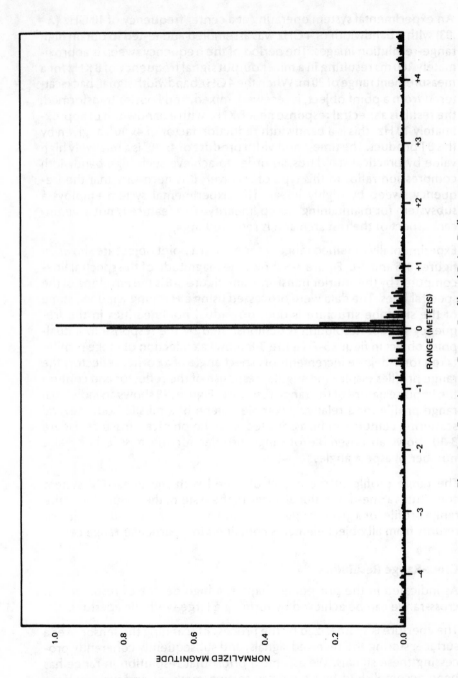

Figure 3-5. Experiment Range Profile of Point Reflector.

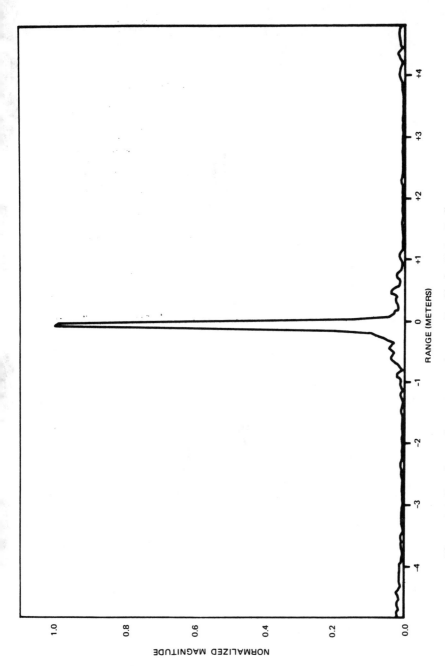

Figure 3-6. Envelope of Experimental Range Profile for Point Reflector.

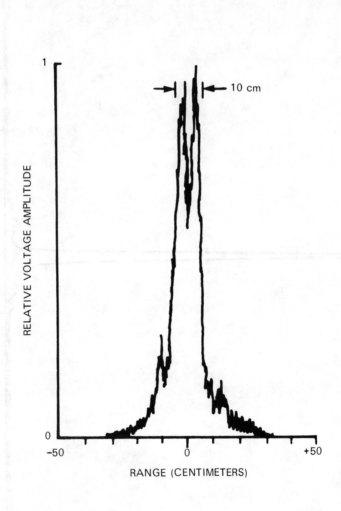

Figure 3-7. Experimental Response of Dual Point Objects Separated by 5 cm.

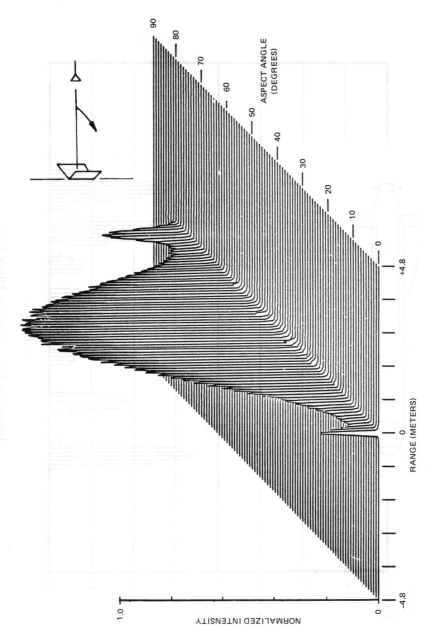

Figure 3-8. Ensemble of Experimental Range Profiles of Corner Reflector.

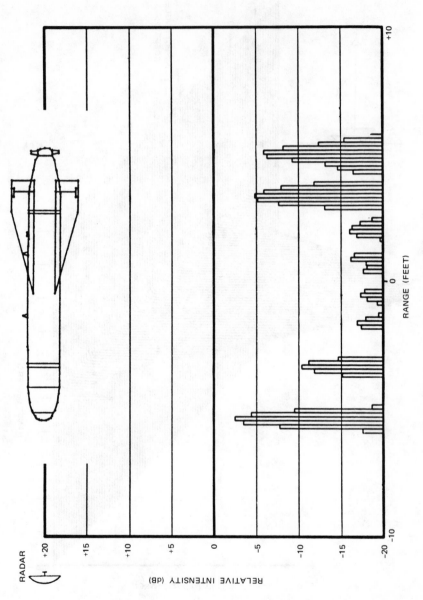

Figure 3-9. Range Profile of Missile Body.

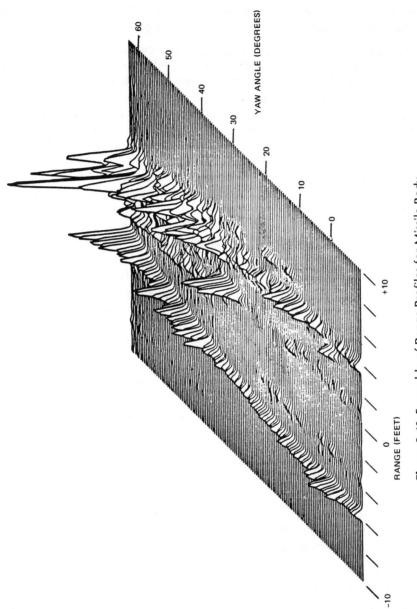

Figure 3-10. Ensemble of Range Profiles for Missile Body.

deal with signals backscattered from sections or slices of the object space contained within a specific range-resolution cell.

When large apertures are required, a significant practical advantage results from forming the synthetic aperture by holding the sensor fixed and rotating the object through a range of angles equal to that which would be subtended by the aperture. The equivalence of observing a rotated object with a fixed sensor is illustrated in figures 3-11 and 3-12. In figure 3-11, the object described in the object-fixed coordinates (x,y) is viewed from an angular offset θ. In figure 3-12, the object has been rotated through the angle θ, and the sensor is collinear with the initial object axis y. In both cases, the relative aspect angle is identical, and the positions of any arbitrary point in the object coordinate frame, relative to the radar are identical. It should be noted that the equivalence of object rotation to sensor rotation is maintained only if the transmitter and receiver are fixed relative to one another, a condition typically inherent to radar systems.

The synthetic aperture is formed by recording the output of a single, fixed sensor for a number of object angular positions, and subsequently coherently adding the resulting signals. The process of rotating the object generates an equivalent circular arc aperture centered in the object center of rotation which thus becomes the focal point of the aperture.

The operations performed are summarized as follows:

1. A microwave transducer, with sufficient angular beamwidth to illuminate the object uniformly, is held fixed and is used to transmit and receive microwave signals.

2. The object is rotated through an array of angular positions, and complex received signals (phase and magnitude), corresponding to each position, are recorded.

3. The signals are processed to effect steering of the synthetic aperture beam across the object. Pre- and post-processing corrections for errors induced by the measurement geometry are introduced.

4. The results of signal processing are displayed to provide a visual image of the object.

The following paragraphs present analyses of the first-order limitations of the rotating-object imaging. The synthetic aperture formation, the processing required for beam steering, the limit of resolution, and the focusing errors are considered.

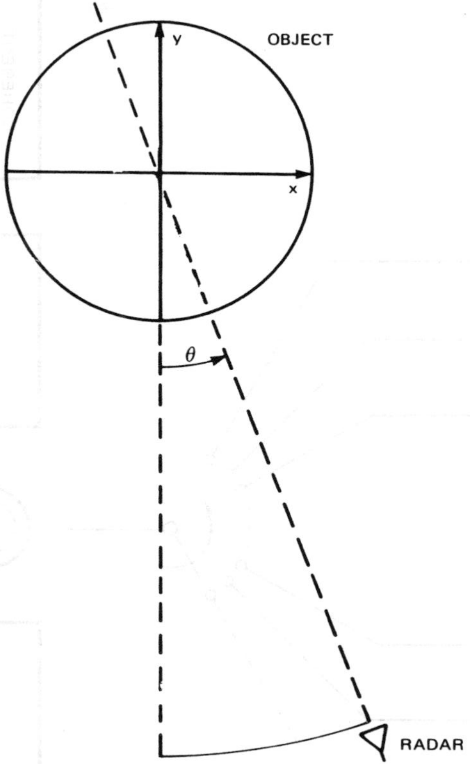

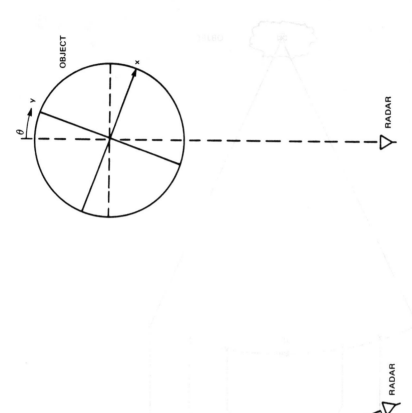

Figure 3-12. Rotated Object Viewed From a Fixed Radar.

Figure 3-11. Fixed Object Viewed From a Radar Offset in Angle.

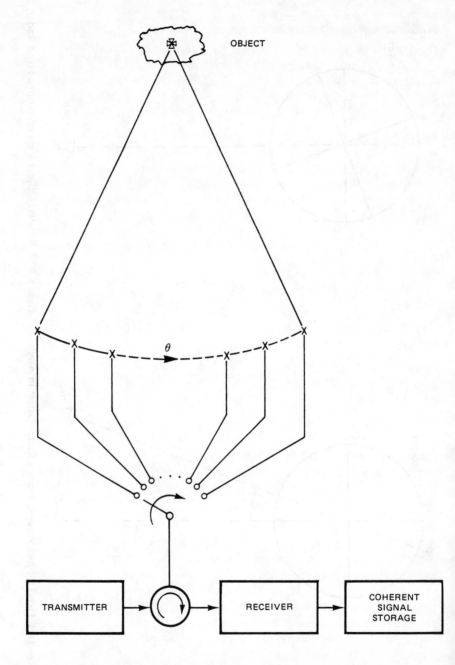

Figure 3-13. Conceptual Synthetic Aperture Forming Process.

Synthetic Aperture Formation

The synthetic aperture is formed by coherently summing received signals which are sampled at sequential angular positions, as diagrammed in figure 3-13. The coherent signal summation is the essential feature of the synthetic aperture formation; this requires that the signals be available in complex (magnitude and phase) form.

Because the sensor is sequentially moved along the aperture plane to obtain samples of backscattered signals, some form of storage device is required to retain the complex sampled signals for subsequent summing. In synthetic aperture processing applied to terrain mapping, the signal information is encoded on film in a manner analogous to a hologram (5) - (7). In our application, the coherent data are sampled, digitized, and stored on magnetic tape for input to a computer memory. Details of the instrumentation are presented in a subsequent section.

The aperture response corresponding to the process in figure 3-13 is represented by the sum:

$$G = \sum_{\theta} V(\theta)$$

where G = the aperture response
   $V(\theta)$ = the complex signal observed at the angle θ.

The signal $V(\theta)$ is the complex envelope of the received signal, represented by the magnitude and phase of the received signal relative to the transmitted signal. The sum of signals received along the aperture is maximum when all signals samples are in-phase. This condition prevails for a point reflector located at the center of rotation, but the phases of signals backscattered from a point in any other location will vary across the aperture thus reducing the magnitude of the summed output.

We consider next the processing required for steering the aperture beam, that is, for directing the peak response of the aperture to a specific point on the x axis. In order to steer the beam to an arbitrary x coordinate, a phase correction must be applied to each sampled field value prior to the summation. The required phase corrections are those necessary to force the summed signals to an in-phase condition. To derive the required correction, consider the geometry of figure 3-14 in which the complex amplitude of the field backscattered from a point object on the x axis is sampled along a circular aperture with radius R_0, where $R_0 \gg x_{max}$. If the collocated transmitter and receiver are moved through an angle θ, the phase of the signal will be advanced due to the shortened length. The received signal phase corresponding to the distance r is:

$$\phi = -4\pi r/\lambda = (-4\pi/\lambda)[R_0^2 - 2xR_0 \sin \theta + x^2]^{1/2} \qquad (3\text{-}9)$$

$$\cong (-4\pi/\lambda)[R_0 - x \sin \theta + x^2/2R_0]$$

The above approximation is obtained from a binomial expansion of the bracketed term, and omitting high order terms using the condition $R_0 \gg x_{max}$. The sampled field values are forced to be in phase if the above phase is subtracted from the sampled values prior to coherent summation. Let $V_x(\theta)$ be the complex field sampled at the angle θ; the response of the synthetic aperture steered to the point x is:

$$G(x) = \sum_{\theta} V_x(\theta) \exp[-j(4\pi/\lambda)(x \sin \theta - x^2/2R_0 - R_0)] \qquad (3\text{-}10)$$

Factoring phase terms not involving θ outside the summation and defining u = sin θ allows rewriting equation 3-10 as:

$$G(2x/\lambda) = \exp[j(4\pi/\lambda)(R_0 + x^2/2 R_0)] \sum_{u} V_x(u) \exp(-j2\pi u \, 2x/\lambda) \qquad (3\text{-}11)$$

where G and V_x are suitably redefined to permit the new arguments. If the sum is performed over uniform increments of u, it represents a DFT where u and $(2x/\lambda)$ are independent variables of a Fourier transform pair.

The term outside the summation is a complex constant that does not affect the magnitude or intensity of the response. If the synthetic aperture is formed over a small angle (u $<$ $\pi/6$) the small angle approximation u $\cong$ θ permits the beam to be steered by a DFT applied directly to data sampled in uniform angular increments. The DFT provides a computationally efficient algorithm for determining the response of the aperture steered over the entire range of x. The steering process is inherent to the DFT which applies to the aperture a phase term linear in θ, and proportional to the coordinate to which the aperture is steered.

The aperture beam can be steered to any value of x, each requiring a distinct summation. If an FFT is utilized for computation speed, the number of samples in the output transform is equal to the number of data samples. Thus, if N data samples are transformed, the aperture is steered to N different values of x. The propriety of the transform operation requires sampling consistent with the Nyquist criterion.

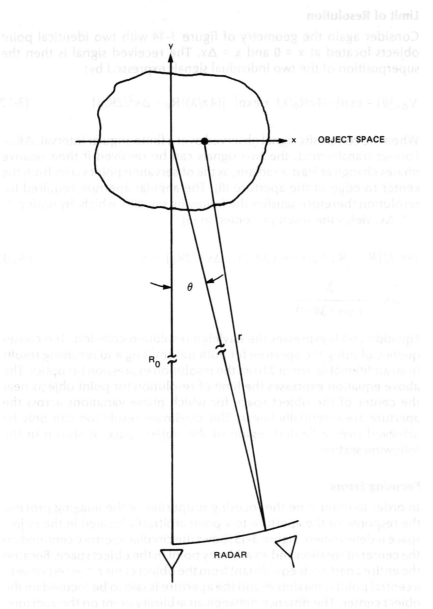

Figure 3-14. Geometry for Sampling Received Field Along a Circular Aperture.

Limit of Resolution

Consider again the geometry of figure 3-14 with two identical point objects located at $x = 0$ and $x = \Delta x$. The received signal is then the superposition of the two individual signals expressed by:

$$V_{\Delta x}(\theta) = \exp[-j4\pi R_0/\lambda] + \exp[-j(4\pi/\lambda)(R_0 - \Delta x^2/2R_0)] \qquad (3\text{-}12)$$

When the composite signal observed over a finite angular interval, $\Delta\theta$, is Fourier transformed, the two signals can be resolved if their relative phases change at least π radians, as the observation point varies from the center to edge of the aperture (8). The angular aperture required for resolution therefore satisfies the following equality which, by noting $R_0 \gg \Delta x$, yields the result presented in (5):

$$(4\pi/\lambda)[R_0 - R_0 + \Delta x \sin (\Delta\theta/2) - \Delta x^2/2R_0] = \pi \qquad (3\text{-}13)$$

$$\Delta x = \frac{\lambda}{4 \sin (\Delta\theta/2)}$$

Equation (3-13) expresses the Rayleigh resolution criterion. The consequence of using the aperture for both transmitting and receiving results in an additional factor of 2 from the resolution expression for optics. The above equation expresses the limit of resolution for point objects near the center of the object space for which phase variations across the aperture are essentially linear. This maximum resolution can only be achieved over a limited region of the object space as shown in the following section

Focusing Errors

In order to determine the focusing properties of the imaging process, the response of the aperture to a point arbitrarily located in the object space is determined. Figure 3-15 shows the circular aperture centered on the center of rotation and an arbitrary point in the object space. Because the entire aperture is equidistant from the object center, the response to a central point is maximum and the aperture is said to be focused on the object center. The distance between an arbitrary point on the aperture, identified by the angle θ, and an arbitrary point in the object space (x,y), is :

$$r(\theta,x,y) = R_0[1 + (2/R_0)(y \cos \theta - x \sin \theta) + (x^2 + y^2)/R_0^2]^{1/2} \qquad (3\text{-}14)$$

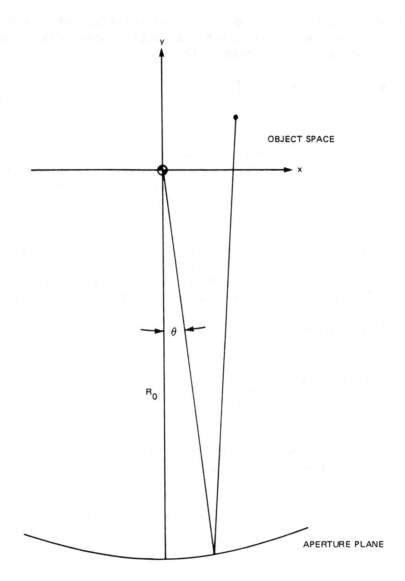

Figure 3-15. Circular Aperture Focused at Object Center.

Field values reflected by a point at (x,y), sampled at the aperture, exhibit a phase angle corresponding to the two-way path length. The normalized amplitude of the received field is:

$$V(\theta, x, y) = \exp\left\{-j4\pi(R_0/\lambda)[1 + 2(y\cos\theta - x\sin\theta)/R_0 + \right.$$
$$\left. (x^2 + y^2)/R_0^2]^{\frac{1}{2}}\right\} \tag{3-15}$$

Equation (3-15) expresses the phase variations observed across the aperture. When signals sampled across the aperture over θ are coherently summed to synthesize the response, the resultant magnitude is reduced from the resultant of a similar sum for an object point at $x = y = 0$. The coherent summation of signals expressed by equation (3-15) for a point-object at (x,y), represents the response of the aperture focused to $x = y = 0$ to a point object located at (x,y). If a phase term equal to the argument of the exponential of (3-15) is subtracted from each signal sample prior to summation, the phase variations will be corrected and the summed signals will be in-phase. The result of the summation will then be a maximum; this operation provides diffraction-limited focusing of the aperture. The response of the focused synthetic aperture to a point object at (x,y) is thus:

$$G(x, y) = \sum_{\theta} V(\theta, x, y) \exp\left\{ j4\pi(R_0/\lambda)[1 + 2(y\cos\theta - x\sin\theta)/R_0 + \right.$$
$$\left. (x^2 + y^2)/R_0^2]^{\frac{1}{2}}\right\} \tag{3-16}$$

The focusing operation constitutes a significant computational burden; equation (3-16) indicates that a distinct phase correction must be applied for each combination of θ and (x,y). If the argument of the exponential term in equation (3-16) is expanded in a series with terms of order 2 and higher neglected, and the following approximations are made,

$$R_0 \gg x, y \qquad \sin\theta \cong \theta \qquad \cos\theta \cong 1$$

equation (3-16) reduces to:

$$G(x, y) \cong \exp[j4\pi(R_0 + y)/\lambda] \sum_{\theta} V(\theta, x, y) \exp(-j4\pi x\theta/\lambda) \tag{3-17}$$

The phase term outside the summation has no effect on the resultant magnitude which can be expressed as:

$$|G(x,y)| \cong |\sum_{\theta} V(\theta,x,y) \exp(-j2\pi\theta\ 2x/\lambda)| \qquad (3\text{-}18)$$

Equation (3-18) represents DFT sum, indicating that an approximate image can be obtained by Fourier transforming the field values sampled along the aperture. The assumptions required for the DFT approximation are that the observation distance be large relative to the object dimensions, and that the angle subtended by the aperture be small. The DFT allows a significant reduction in the computational load, offering the feasibility of near real-time imaging in some applications.

The development of equations (3-18) and (3-14) indicates that, in the limit of large observation distances and small angular aperture, the DFT is a valid approximation for both steering and focusing.

Using equation (3-16) to process sampled field values steers the synthetic aperture beam to any point on the object plane and produces diffraction-limited focusing; using equation (3-18), the DFT approximation steers the aperture beam which remains focused to the diffraction limit at the object center, but becomes defocused when directed to other points in the object plane.

The nature of the focusing error can be determined as follows. When sampled field values are processed with a DFT, a sequence of phase corrections linear in θ is applied; although this has the effect of steering the aperture in a general direction determined by the particular phase slope, it does not provide the phase correction necessary to focus the aperture exactly at any particular point. As a result, the response of the aperture thus synthesized is decreased from that of focused aperture. Figure 3-16 shows the equivalent geometries of the synthetic apertures formed by the various processes. The unsteered aperture, inherently focused to the object center, is represented by sampled locations along the circular arc centered on the object center. The steered focused aperture, formed by applying a phase correction to the sampled field values, such as to bring them to an in-phase condition, is a circular arc centered on the coordinates (x_0,y_0) to which the aperture is focused. The differential path length representing the required phase correction is denoted by Δ. The DFT-steered aperture is formed by applying to the sampled field values discrete phase corrections proportional to x and θ. This quantization in beam steering precludes tangency between the

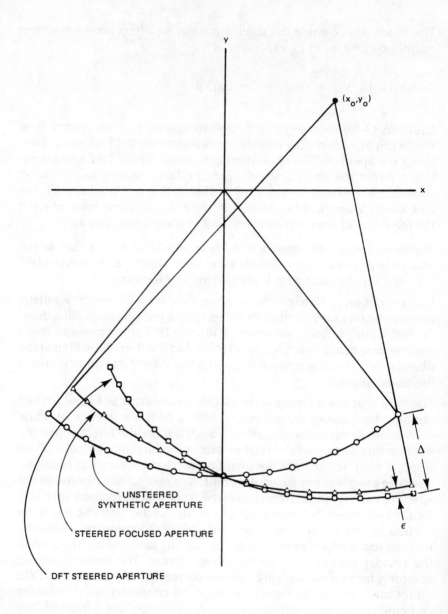

Figure 3-16. Geometry of Synthetic Apertures.

DFT-steered and steered focused apertures, and results in phase errors corresponding to the differential path length denoted by ε. Processing the data with an N-point DFT is equivalent to generating N steered beams, each formed by a phase correction proportional to x. Each of the N beams is only approximately focused in the vicinity of the x coordinate due to the phase errors inherent to the DFT approximation.

To develop the response of the steered aperture, we consider the sampled aperture geometry of figure 3-17. The phase of the signal reflected from a point object located at x_0, y_0, and received at an aperture coordinate θ_n is:

$$\phi(x_0, y_0; \theta_n) = \frac{-4\pi}{\lambda} \left\{ [R_0^2 + 2R_0(y_0 \cos \theta_n - x_0 \sin \theta_n) + x_0^2 + y_0^2]^{\frac{1}{2}} \right\}$$

(3-19)

The DFT of the sequence of signals observed across the aperture is represented by:*

$$G(x_0, y_0; x_k) = \sum_n \exp[j\phi(x_0, y_0; \theta_n)] \exp(-j2\pi \, 2x_k\theta_n/\lambda)$$
(3-20)

where $\phi(x_0, y_0; \theta_n)$ is defined in equation (3-19).

Equation (3-20) is the response to a point reflector located at x_0, y_0 of the aperture steered to x_k. The form of the response of the steered beams can be determined more easily by allowing the observation distance, R_0, to approach infinity. Rewriting the bracketed terms of equation (3-19) in a binomial expansion and assuming $R_0 \gg x_0 + y_0$ provides an expression for the phase variation across the aperture in the limit of infinite observation distance. Equation (3-20) can then be rewritten as:

$$G_\infty(x_0, y_0; x_k) = \sum_n \exp \left\{ -j4\pi(R_0/\lambda)[1 + (y_0 \cos \theta_n - x_0 \sin \theta_n)/R_0 \right.$$

$$\left. + (x_0^2 + y_0^2)/2R_0^2] - [j2\pi\theta_n \, 2x_k/\lambda] \right\}$$

(3-21)

The above expression can be rewritten by factoring terms not involving θ

---

*For this development the amplitude variations of the observed signal are ignored and a point reflector is assumed to provide unit-magnitude received signals at all aperture element positions. This approximation is consistent with the far-field assumption.

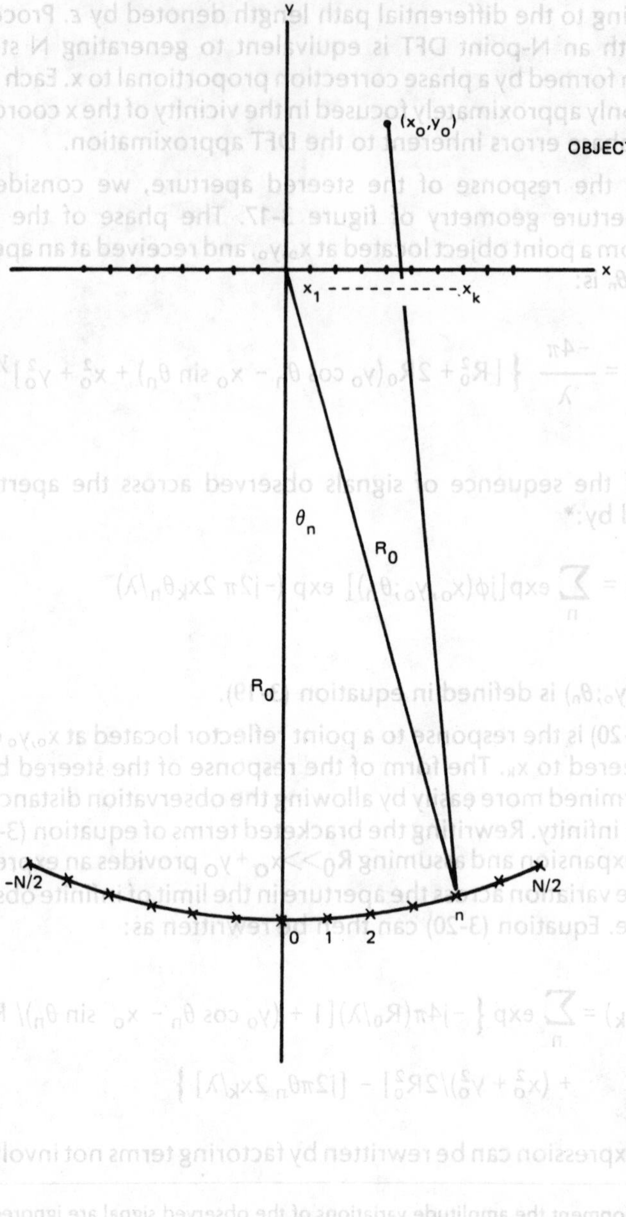

Figure 3-17. Sampled Aperture Focused at Object Center.

outside the summation, yielding:

$$G_\infty(x_o, y_o; x_k) = \exp\left\{-j4\pi[R_0/\lambda + (x_o^2 + y_o^2)/R_0^2]\right\}$$

$$\sum_n \exp\left\{-j4\pi/\lambda[(y_o \cos\theta_n - x_o \sin\theta_n) + x_k\theta_n]\right\} \quad (3\text{-}22)$$

The subscript on G indicates that the response applies to an infinite observation distance; we shall subsequently refer to it as the far field response.

The general form of the response for the central beam ($x_k = 0$) is shown in figure 3-18, which plots the intensity $|G^2|$, given by equation (3-22) for $x_k = 0$ and a total aperture angle of 30 degrees. The contours are lines of constant intensity for uniform increments of 0.1 times the peak value. The peak response of the unsteered beam occurs for $x_o = y_o = 0$, because signals reflected from this point and received across the aperture are in phase. As the location of the object point deviates from center, phase errors developed across the aperture cause a reduction in the magnitude of the response. The precise shape of the response for each beam is determined by plotting equation (3-22) for each beam. A tractable expression for estimating the size of the main response of the beams can be developed by establishing a phase-error criterion. The response of an aperture without phase error has a maximum amplitude and minimum beamwidth; when the aperture is subject to phase errors, the magnitude is reduced and the beamwidth broadened. The limit of allowable phase errors is somewhat arbitrary but a criterion which is generally accepted (1) allows maximum phase errors of $\pi/2$ radians at the edges of the aperture. If the error is quadratic across the aperture, the peak magnitude of the response is decreased by less than 15 percent and the spatial response widened by less than 10 percent (6).

By adopting the $\pi/2$ radians maximum phase error, we can determine the boundaries of locations of point-objects in the x-y space which result in aperture phase errors equal to the prescribed limit. Points inside these boundaries will be imaged by some degradation; points outside these boundaries will be imaged with a magnitude reduction determined by the deviation from the boundary. The phase of signals reflected from a point-object located at x_o, y_o and received at the end and center of the steered aperture, are given respectively by the exponent of equation (3-21) for $\theta_n = \Theta/2$ and $\theta_n = 0$. The phase difference is then:

$$\frac{4\pi}{\lambda}\left\{-y_o\left[1 - \cos\left(\frac{\Theta}{2}\right)\right] - x_o \sin\left(\frac{\Theta}{2}\right) + x_k\left(\frac{\Theta}{2}\right)\right\} = \frac{\pi}{2} \tag{3-23}$$

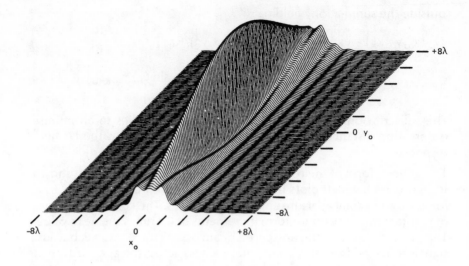

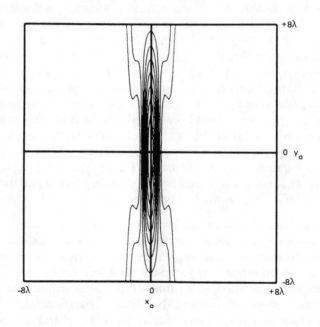

Figure 3-18. Intensity of Far-Field Response for
Unsteered Beam of 30-Degree Aperture.

where Θ is the total angle subtended by the aperture for which a maximum phase error of $\pi/2$ radians can occur at either edge of the aperture

Figure 3-19 shows the focal zone pattern, defined by equation (3-23), when the N sampled data are processed by an N-point DFT to produce N steered beams. The focal zones are contiguous without overlap. The zones can be made to overlap by computing additional beams, accomplished by padding the sampled data with zeros and computing an M-poing DFT where M>N. The following observations can be made from figure 3-19: (1) the depth (y extent) of the focal zones remains unchanged with x_k and (2) the peak response for a beam steered to x_k occurs for a point object location x_0 such that $|x_0| > |x_k|$. The maximum response of a particular beam occurs for a location slightly offset from that to which the beam is steered; the offset, which increases with x_k, results from performing the DFT using equal increments in θ, rather than in $u = \sin \theta$. This limitation could be removed at the cost of increased instrumentation or computational complexity. The lateral extent of the object space in which points can be imaged can be defined somewhat arbitrarily by the condition of having the x_k coordinate, to which the beam is steered, fall on the boundary of the focal zone of the kth beam. This condition is shown in figure 3-20 in which the overall extent of the object field is shown by the large rhombus. Responses for beams of a 30-degree aperture steered to values of x_k denoted A through E in figure 3-20 are shown in figures 3-21 through 3-25, obtained by using equation (3-21). The contours are again lines of constant intensity with increments 0.1 times the peak value of the unsteered beam. The offset in the response of the steered beams can be noted in these figures. Figures 3-26 and 3-27 show responses for the aperture steered to an x_k coordinate 2 times and 4 times that for point E, respectively. These figures illustrate the severe degradations for beams steered to large lateral offsets.

In the preceding analysis, the allowable degradation has been established by the arbitrary selection of $\pi/2$ radians maximum phase error. This is relatively conservative because it limits the degradation to a 15% reduction in intensity. The general dependence of the focal zones on the aperture angle is valid for aperture angles less than 1 radian. For larger apertures, the phase errors become excessive and the $\pi/2$ radian error criterion is no longer valid. For the majority of applications, this restriction is satisfied and the dimensions of the object field can be quickly estimated for a given aperture angle, using the relations shown in figure 3-21. In the limit of small aperture angles, the lateral and longitudinal

HIGH RESOLUTION RADAR IMAGING

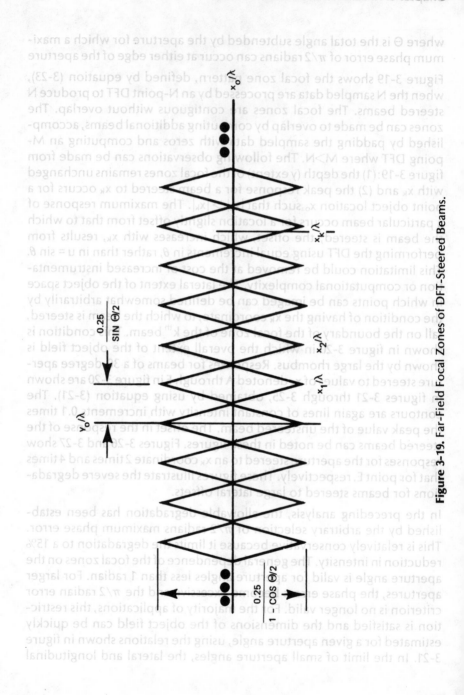

Figure 3-19. Far-Field Focal Zones of DFT-Steered Beams.

extent of the focal zone for the central beam become respectively:

$$\Delta x = \frac{0.25\lambda}{\sin(\Theta/2)} \cong \frac{\lambda}{2\Theta} \tag{3-24}$$

$$\Delta y = \frac{0.25\lambda}{1 - \cos(\Theta/2)} \cong \frac{0.25\lambda}{1 - [1 - (\Theta^2/4)]} = \frac{\lambda}{\Theta^2}$$

Noting the following inequality between the lateral and longitudinal field extents cited in figure 3-20,

$$\frac{0.125}{(\Theta/2) - \sin(\Theta/2)} \geqslant \frac{0.125}{1 - \cos(\Theta/2)} \tag{3-25}$$

we conclude that the maximum radial distance of a point-object that can be imaged with tolerable degradation is limited by the depth of field established by the longitudinal extent of the focal zone. Because the lateral (cross-range) resolution and the depth of field both depend on the aperture angle, a fundamental relation between lateral resolution and maximum object dimension can be established. The ratio of field depth to the width of the lateral resolution cell is given by the aspect ratio of the focal zones in figure 3-20. Let x be the lateral resolution cell size; then, from figure 3-20:

$$\frac{\Delta x}{y} = \frac{0.25}{\sin(\Theta/2)} \tag{3-26}$$

The number of lateral resolution cells contained in the field-depth extent is:

$$N = \frac{\sin(\Theta/2)}{1 - \cos(\Theta/2)} \tag{3-27}$$

The total extent represented by N cells of dimension Δx is:

$$D = N\Delta x = \frac{0.25\lambda}{1 - \cos(\Theta/2)} \tag{3-28}$$

Combining the above three equations to eliminate Θ yields;

$$\frac{D}{\lambda} = \frac{0.25}{1 - \cos\left[\sin^{-1}\left(\frac{0.25\lambda}{\Delta x}\right)\right]} \qquad N = \frac{0.25}{\frac{\Delta x}{\lambda}\left\{1 - \cos\left[\sin^{-1}\left(\frac{0.25\lambda}{\Delta x}\right)\right]\right\}} \tag{3-29}$$

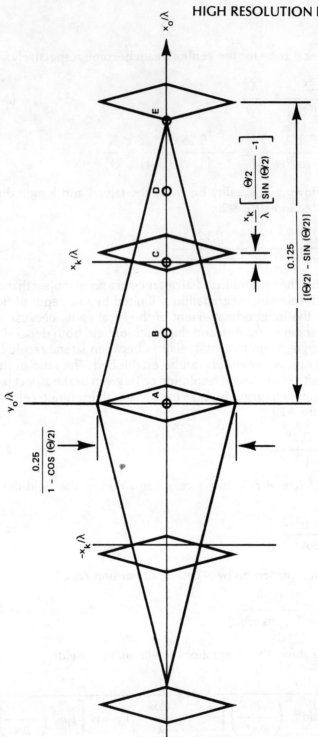

Figure 3-20. Spatial Extent of Object Field for DFT-Steered Aperture.

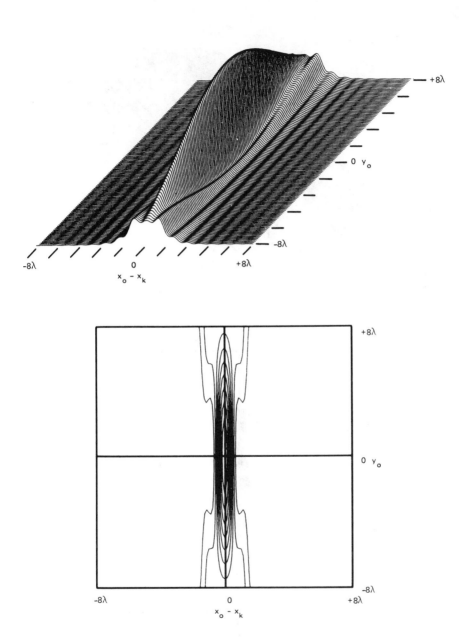

Figure 3-21. Response of 30-Degree Aperture Beam Steered to Point A.

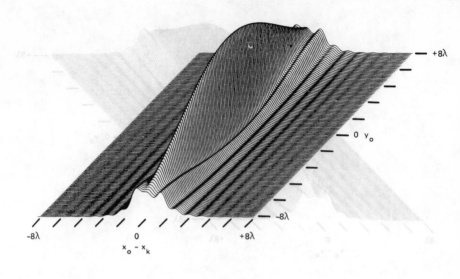

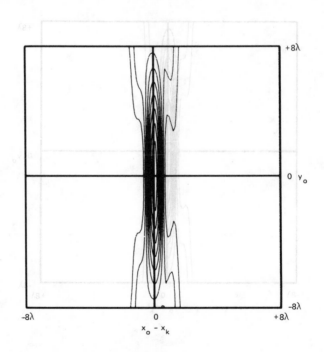

Figure 3-22. Response of 30-Degree Aperture Beam Steered to Point B.

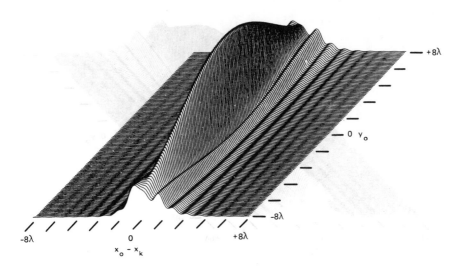

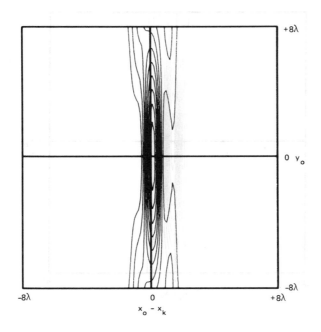

Figure 3-23. Response of 30-Degree Aperture Beam Steered to Point C.

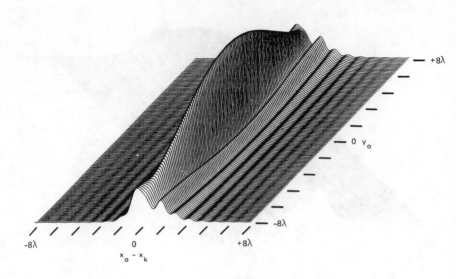

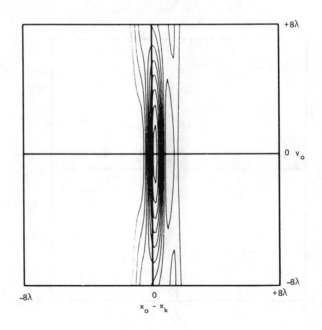

Figure 3-24. Response of 30-Degree Aperture Beam Steered to Point D.

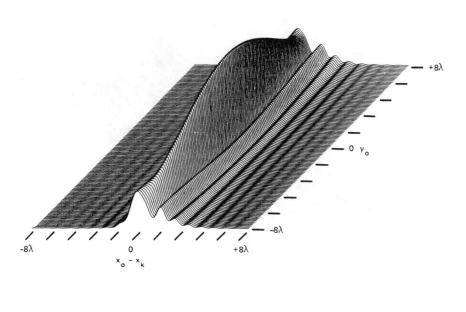

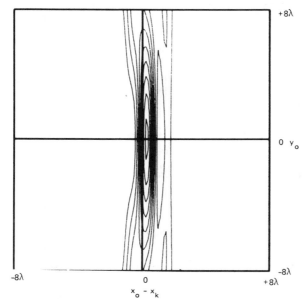

Figure 3-25. Response of 30-Degree Aperture Beam Steered to Point E.

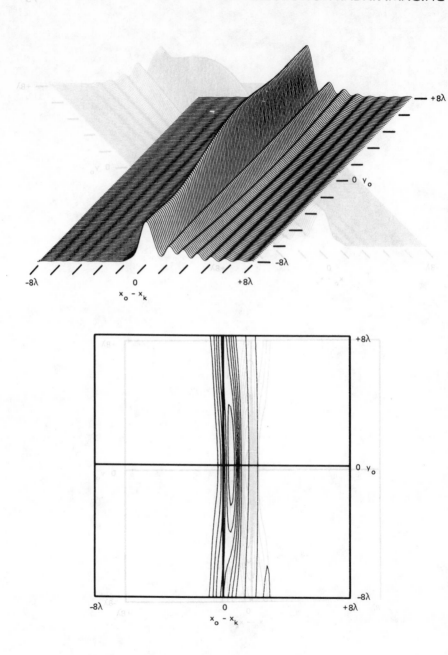

Figure 3-26. Response of 30-Degree Aperture Beam Steered to Point 2E.

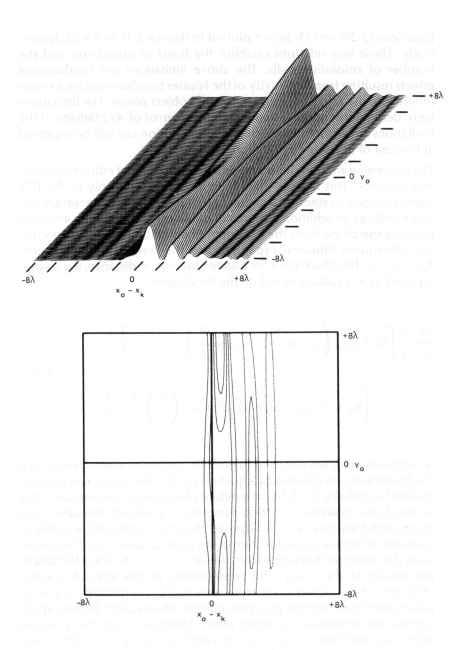

Figure 3-27. Response of 30-Degree Aperture Beam Steered to Point 4E.

Equations (3-29) and (3-30) are plotted in figures 3-28 and 3-29, respectively. These two relations establish the limits of object size and the number of resolution cells. The above limitations are fundamental effects resulting from sensitivity of the Fourier transform to phase errors inherent to the circular motion of rotating object points. The limitations were derived by adopting a tolerable phase error of $\pi/2$ radians; if the limitations are violated, some degree of resolution can still be expected at the cost of imaging degradation.

The preceding analysis was based on the assumption of infinite observation distance. This implies that phase errors are due only to the DFT approximation to steering a circular aperture. A finite observation distance induces an additional source of phase error that can be analyzed by using the phase from the exponent of equation (3-20) and omitting the subsequent binomial expansion and the simplifying assumption, $R_0 >> x_0 + y_0$. The phase error for a finite observation distance, R_0, is again equated to $\pi/2$ radians to define the focal zones.

$$\frac{4\pi}{\lambda} \left\{ \left[R_0^2 + 2R_0 \left(y_0 \cos \frac{\Theta}{2} \mp x_0 \sin \frac{\Theta}{2} \right) + x_0^2 + y_0^2 \right]^{\frac{1}{2}} \right.$$

$$\left. - \left[R_0^2 + 2R_0 y_0 + x_0^2 + y_0^2 \right]^{\frac{1}{2}} \pm x_k \left(\frac{\Theta}{2} \right) \right\} = \frac{\pi}{2} \tag{3-31}$$

A computer program was used to perform a numerical search to define the boundaries established by equation (3-31). The results can be summarized as follows: Each beam exhibits a focal zone very similar to the far-field case, however, the focal zones are aligned with the axes of the beams which are now radially disposed about the center of the synthetic aperture (which is also the location of the physical aperture). This constitutes the principal difference from the far field case in which the beams are parallel to the y-axis. The consequence of this effect is a radial distortion of the object space from rectangular coordinates to a set of polar coordinates with origin at the point of observation. In most applications, this distortion is a minor effect. However, it can be corrected after range and cross-range imaging is performed. Figures 3-30 through 3-34 show plots of near-field focal zones. The conditions are identical to those used for the far field responses shown in figures 3-21 through 3-25, but the observation distance R_0, was 25 m. and the wavelength 0.03 m.

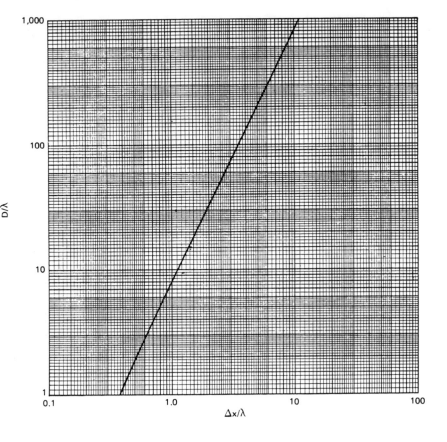

Figure 3-28. Maximum Object Field, D, Versus Lateral Resolution, Δx.

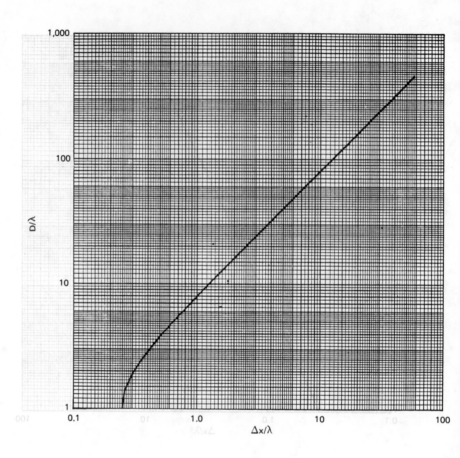

Figure 3-29. Maximum Number of Resolution Cells, N, Versus Lateral Resolution, Δx.

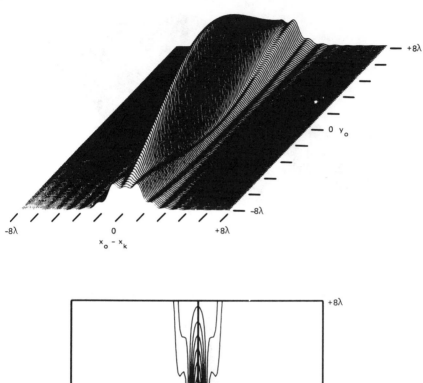

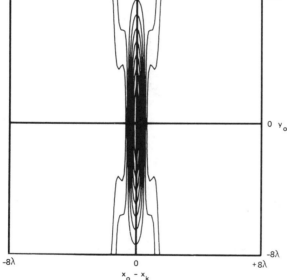

Figure 3-30. Near Field Response of 30-Degree Aperture Beam Steered to Point A.

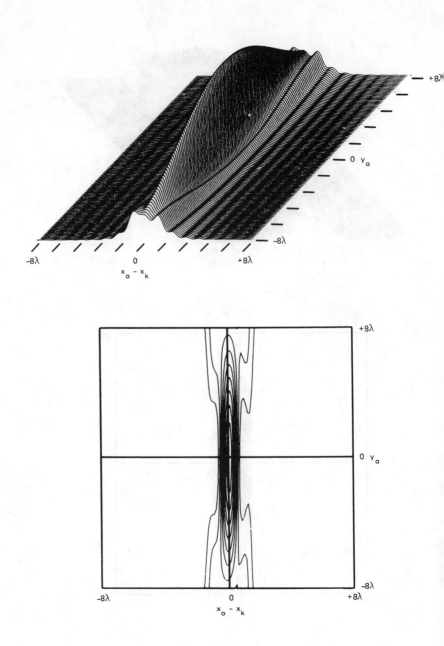

Figure 3-31. Near Field Response of 30-Degree Aperture Beam Steered to Point B.

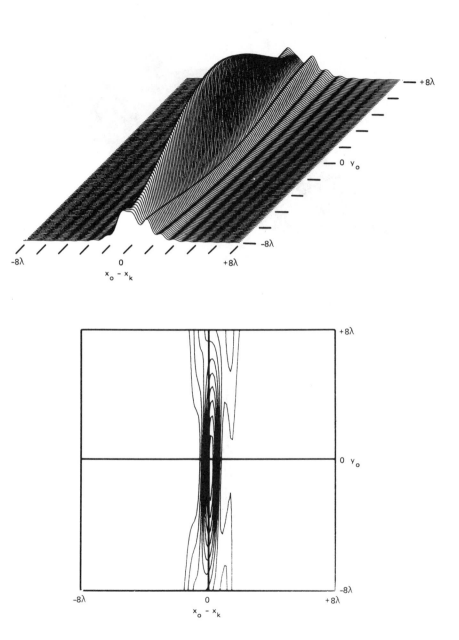

Figure 3-32. Near Field Response of 30-Degree Aperture Beam Steered to Point C.

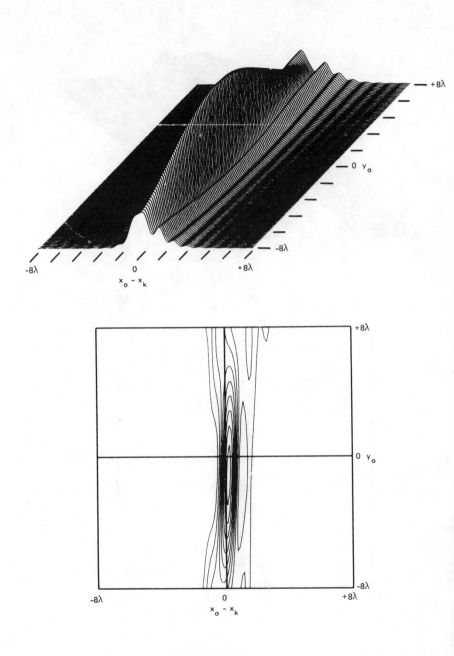

Figure 3-33. Near Field Response of 30-Degree Aperture Beam Steered to Point D.

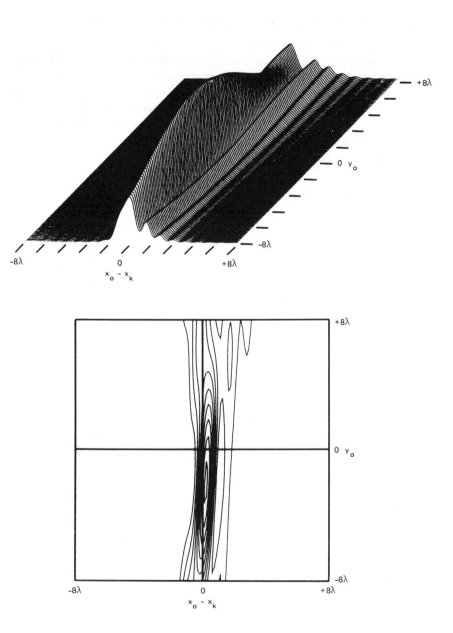

Figure 3-34. Near Field Response of 30-Degree Aperture Beam Steered to Point E.

Two-Dimensional Imaging in Range and Cross-Range

We now describe parameters of a system used for collecting experimental data and present examples of images obtained from the data. Because the implementation of the equipment is not central to the discussion, specific details of the hardware are omitted and only aspects pertinent to the signal processing are discussed. A block diagram of the experimental setup is shown in figure 3-35. System parameters have been selected to provide equal resolution in range and cross-range, and adjusted to accommodate sample numbers which are powers of 2, in order to exploit FFT algorithms.

For the examples presented here, the microwave system was operated with a center frequency of 10 GHz and a bandwidth of 1.57 GHz. Linear modulation of the transmitted signal frequency was triggered by an angular encoder connected to the object rotator; the chirp period was 51.2 ms. The low-frequency product of transmitted and received signals, nominally a 5 KHz tone, was sampled at a 20 KHz rate, converted to six-bit binary words, and recorded. The sampling rate was sufficient to avoid aliasing errors and produced 1024 samples of data during the chirp period. After the sampled data were recorded, the object was rotated to the next angular position and the process was repeated until the object had undergone a complete rotation. Each block of 1024 data samples was multiplied by a cosine-squared weighting (Hanning window (3)) to reduce range sidelobes and was padded with 1024 zeros, as shown in figure 3-36. The data were processed by a 2048-point FFT to produce 2048 spectral samples, of which 1024 were unambiguous. The frequency index corresponding to the center of the object range field was identified, and 64 complex spectral samples (real and imaginary) on each side of center retained. The 128 spectral samples represent a range space of 6 m.

The angular increment between data blocks was 0.14 degrees which satisfied the sampling criterion for signals from scatterers with a maximum cross range of 3m. Sixty-four such angular increments constitute the 9-degree angular extent of the aperture. The 64 individual spectra were placed in columns of a two-dimensional array. Each row was multiplied by a cosine-squared weighting to reduce cross-range sidelobes in the subsequent FFT and the array was padded with 64 additional rows containing zeros. The configuration of the array is shown in figure 3-37. The final operation was a 128-point FFT applied to the data (magnitude and phase of the range profiles) in each of the 128 rows.

The structure of the array in figure 3-37 is summarized as follows. The samples forming each column are obtained by Fourier transforming the

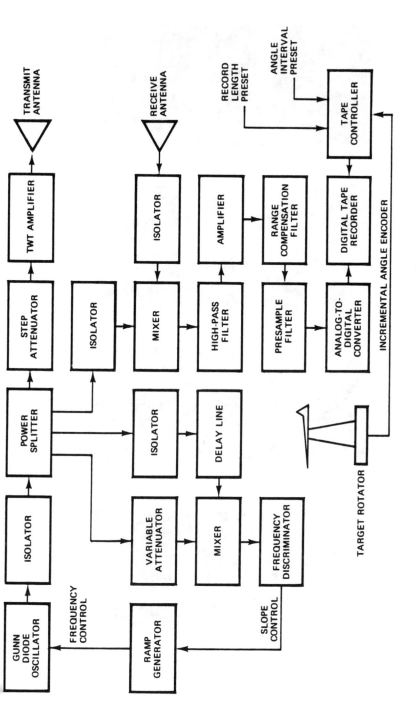

Figure 3-35. Block Diagram of Experimental Systems.

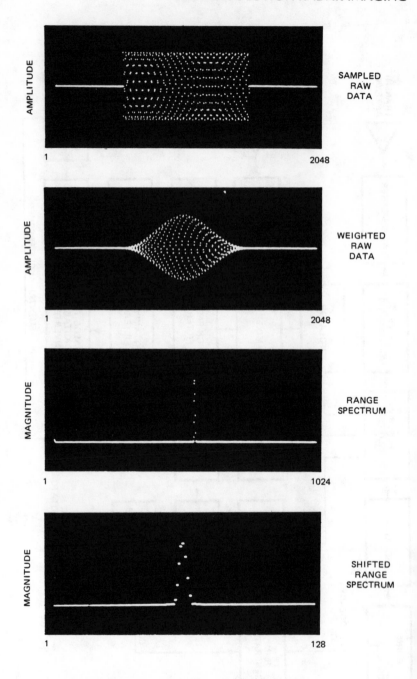

Figure 3-36. Structure of Data for Range Processing.

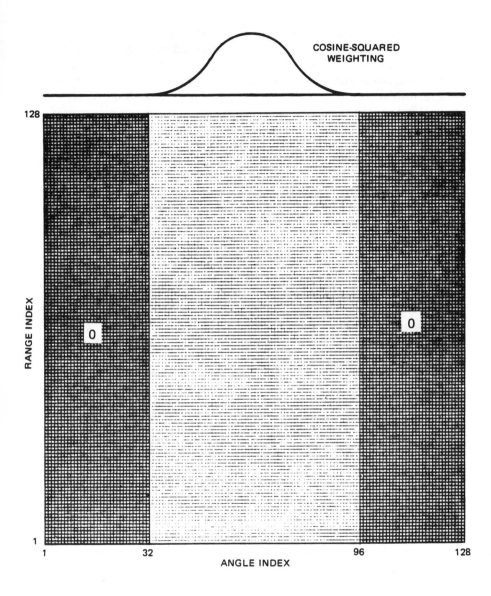

Figure 3-37. Configuration of Data Array for Cross-Range Processing.

real input data and they represent the magnitude and phase of the sum of signals reflected from all points contained in each range cell. Each row, in turn, represents the variation in the content of a particular range cell. Fourier transforming the content of each row steers the synthetic aperture to 128 different locations, and thus describes a cross-range profile for each range cell. The results is a two-dimensional array which constitutes the range and cross-range of the object's reflectivity density. The weighting applied prior to both transforms suppresses the sidelobes in both range and cross-range and the padding with zeros enhances the spectral definition. The process maps the reflectivity distribution of the object in a 6m by 6m square field with equal resolution in range and cross-range of the order to 20cm. The resolution in the two directions has been set equal by proper selection of parameter values in accordance with equations 3-8 and 3-13.

The computational procedure just described performs a 2048 point FFT of the recorded data and discards all but 128 spectral samples. This is because the object occupies a limited portion of the total range space. The discarding of the unused 1920 samples represents a computational inefficiency which can be overcome by zoom FFT techniques (9) in which the required 128 samples are computed directly. This is accomplished by implementing a digital band-pass filter operating on the 1024 sample input sequence, decimating to 128 samples and computing a 128-point FFT. This operation, properly implemented, provides results identical to the previous procedure. Although the former method is less efficient, it was used to produce the experimental images because it minimized development of new computer programs.

The combination of zoom FFT and increased memory capacity allows a further computational simplification to the process. The input data can be bandpassed and decimated, resulting in 128 data samples which can then be loaded in columns of a two-dimensional array. For each increment of angular rotation, a new column is generated until a 128 x 128 array is formed. A two-dimensional window is then applied to the array and a two-dimensional FFT computed. General purpose computers currently available have sufficiently large memories to allow the direct computation of 128 x 128 point two-dimensional FFTs.

In order to test the computational process, simulated data were generated to represent the idealized signal reflected from a linear array of four independent point objects shown in figure 3-38. The composite received signal is given by the sum of four signals, each expressed by equation (3-5) with the range corresponding to each point. The simulated received signal was computed for 0.14-degree angular increments and processed as an actual signal would be. The intensity of the resulting images is shown in figure 3-39. Each image is obtained by processing data from a 9-degree sector. Three computed images are shown for angular sectors centered on 0, 36, and 90-degrees. The response of the central point represents the point-spread function of the imaging system and conforms well to the theoretical resolution of 20cm. The sidelobes in the point-spread response are adequately suppressed by the windowing.

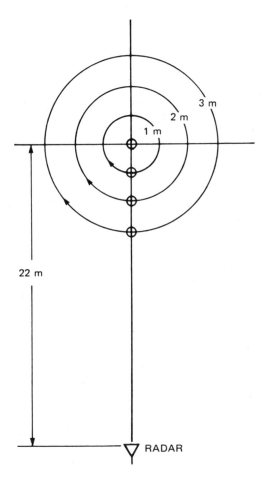

Figure 3-38. Configuration of Linear Array of Four Point Objects.

Responses of points in the other three quadrants are symmetrical to those shown. The point-spread function exhibits some degradation for object points displaced from center. Images of points near the boundaries of the object space are reduced in intensity by a factor of two and the spatial response is broadened. The degradation for points with large offset in range from the center of rotation is due to the focusing error described in the preceding section. Points with maximum cross-range are also imaged with similar degradation because they traverse more than one range-resolution cell in the course of the 9-degree rotation. This effect is reduced by the windowing process which decreases the effective angular extent. A slight artifact is noted at the right of the 90-degree image. This is aliased leakage from the response to the left caused by the periodic nature of the FFT. It is not significant and can be avoided by ensuring the object is well contained in the unambiguous image space. Figure 3-39 represents the imaging performance of the process described. Subject to the degradations noted over the extent of the image field, the method is adequate for two-dimensional imaging of objects and represents the limit of achievable performance with relatively simple standard FFT algorithms.

The effect of windowing the data prior to Fourier transforming is demonstrated by processing the identical data with a rectangular window replacing the cosine-squared window in both (range and cross-range) transforms. The results, shown in figure 3-40, differ from previous results in the following aspects: the resolution of the central point is enhanced, minor sidelobes caused by the abrupt truncation of the rectangular window are evident; and the off-center points are imaged with increased spreading and reduced peak magnitude. This degradation is a result of the increased effective width of the windows. This decreases both the depth of field and the size of the range-resolution cell, thereby causing more smearing than in the previous case. All subsequent images are processed using the cosine-squared window.

Several examples of images processed from experimental data are shown in the following figures. Figure 3-41 is the image of two 10-cm diameter, 15-cm length cylinders oriented with their longitudinal axes parallel to the axis of rotation. The two cylinders are diametrically opposed 120 cm inches from the center of rotation. Due to their small physical size, the cylinders behave essentially as point objects. The image field is free from spurious responses, artifacts and noise. The slight difference in peak intensity is caused by errors in vertical alignment of the cylinder axes.

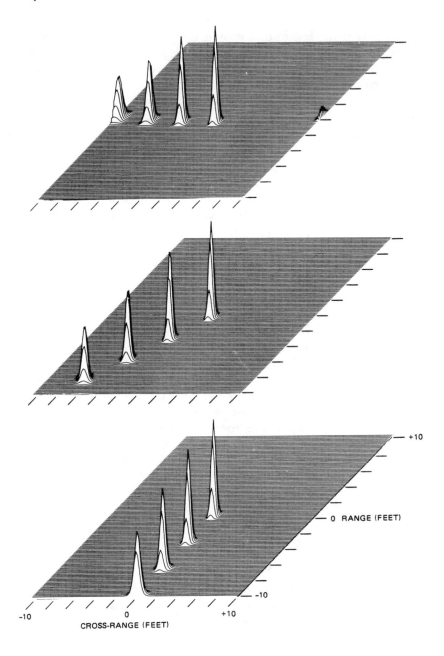

Figure 3-39. Intensity of Image Processed From Simulated Data Using
Hann Windows.

Figure 3-42 is the image of a missile body shown superimposed. The image was obtained by processing the range profiles shown in figures 3-9 and 3-10. This example demonstrates the value of two-dimensional imaging: resolution in cross range allows the separation of the two scatterers at the rear of the missile which would not be separated by range processing alone.

Figures 3-43 and 3-44 are images of a human. These data were obtained by rotating the body supported in the supine position about a vertical axis. The superimposed ellipse indicates the area covered by the rotated body; the inscribed arrow indicates the orientation with the point of the arrow denoting the head position. The arrow outside the ellipse indicates the direction of viewing. The level of microwave exposure for these images was $0.05 \, mw/cm^2$, a value well below the accepted safe level of 10 mw/cm^2. Although the utility of such images has not been established, they nevertheless constitute the first reported measurements of the radar reflectivity density of a human body.

Figure 3-45 is an image of another missile body in the orientation indicated by the diagram. The single response at the far end of the image field is a range marker. The front part of the missile, denoted by the head of the arrow, is imaged with considerable range extent. This is caused by multiple internal reflections of signals penetrating the transparent radome. The multiple reflections constitute time delays which are interpreted in the imaging process as extended range. An additional feature of interest is the minor ridge along the zero cross-range axis which is caused by low-level reflections from the anechoic enclosure used for the measurements. Because stationary objects generate no change in phase as a function of angular rotation of the test object, they do not produce a Doppler shift, and are therefore imaged with zero cross range. Their range position, however, is properly imaged because range imaging is accomplished independently of object rotation. The result in this type of two-dimensional processing is that images of stationary objects are collapsed on the zero cross-range axis.

Extension to Three-Dimensional Imaging

In the preceding section, we described a method for two-dimensional imaging which used time-delay sorting to obtain resolution in range and object rotation for resolution in cross-range along an axis normal to the axis of rotation of the object. This technique is adequate for imaging three-dimensional objects if resolution is obtained in the second cross-range direction by using fan beams to irradiate selected slices of the object. When the entire object is uniformly irradiated, however, the

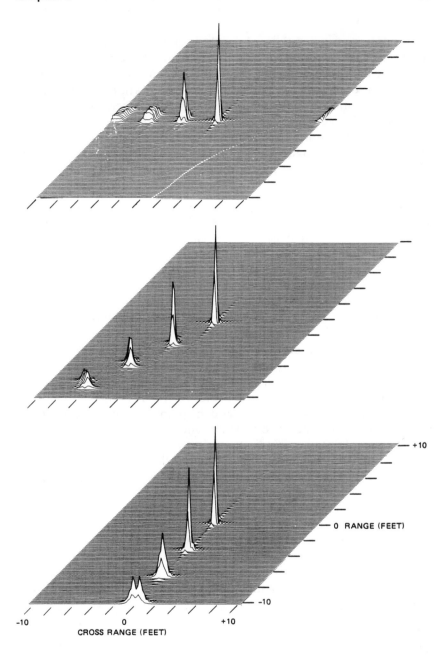

Figure 3-40. Intensity of Image Processed From Simulated Data Using Rectangular Windows.

images represent projections of the three-dimensional reflectivity distribution of the two-dimensional image plane. In some applications, such as the locating of scattering centers on known objects (figures 3-40 to 3-45), the lack of resolution in the third dimension is not critical. In these cases, the location of reflecting points can be inferred from the two-dimensional image because the object shape is known. Thus, a priori information can be used to remove ambiguities not resolved by the imaging process. When imaging complex objects without a priori information of the reflectivity distribution, however, three-dimensional resolution may be required to unambiguously define the image.

In this section we outline a method to obtain resolution in both cross-range directions by rotation of the object about two mutually orthogonal axes. The signal content of each range cell, resolved by the same method used in the preceding section, is coherently processed as a function of the two rotation angles. We again adopt the viewpoint that observing a rotating object with a stationary radar is equivalent to observing a stationary object with a radar moving on a spherical surface. This condition is shown in figure 3-46, in which the object is rotated by the angles θ and ψ about the y- and x-axes, respectively. The total angular excursions, Θ and ψ, form a two-dimensional aperture which is a section of a sphere. For each range cell, the image of the object is found by a two dimensional DFT of the received signals sampled as a function of the two angles θ, ψ. Because the angles Θ and ψ are relatively small ($< \pi/6$), the synthesized aperture is nearly rectangular and the point-spread function, $h(x,y)$, is separable in rectangular coordinates and given by the product $h_x(x)$ $h_y(y)$. As a result, each of the relations derived for the one-dimensional synthetic aperture applies to the two-dimensional case. The required signal processing is more extensive than in the preceding case because a one-dimensional DFT is replaced with a two-dimensional DFT to process the signal content of each range cell as a function of rotation angle. Although more stored data and computations are required for two-dimensional synthetic aperture case than for the one-dimensional case, the types of data and computations are similar; resolution in the second cross-range dimension can thus be achieved at the cost of increased storage capacity and computation time.

The physical implementation of the two-dimensional synthetic aperture requires means for rotating the object about two orthogonal axes. This can be accomplished by a gimballed support system or by using other mechanisms by which the object can be independently tilted and rotated relative to the radar axis. Although these procedures for obtaining three-dimensional resolution are a direct extension of the two-

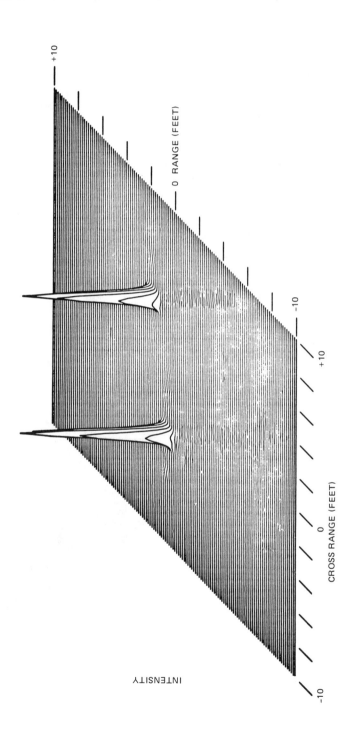

Figure 3-41. Image of Two Cylinders.

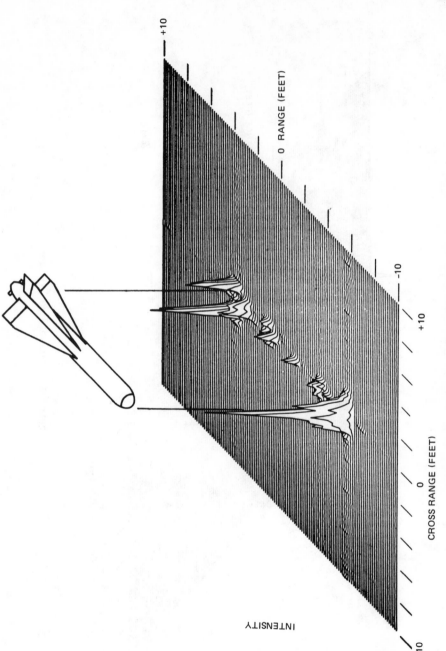

Figure 3-42. Image of Missile Body in 0-Degree Position.

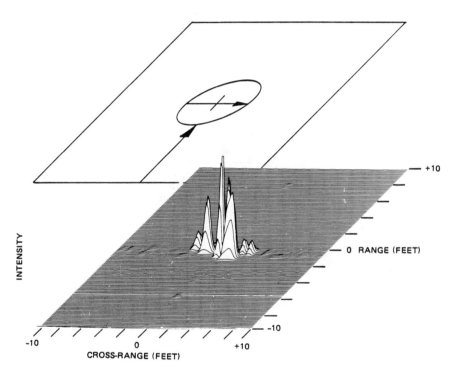

Figure 3-43. Image of Human Body in 90-Degree Position.

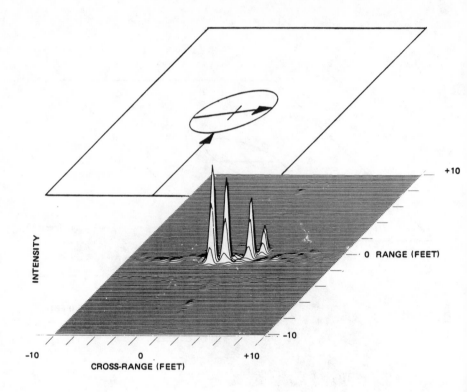

Figure 3-44. Image of Human Body in 108-Degree Position.

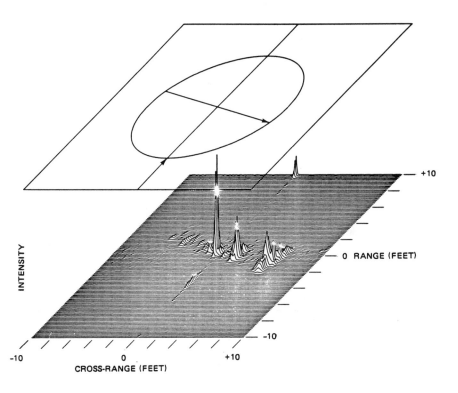

Figure 3-45. Image of Missile Body in 72-Degree Position.

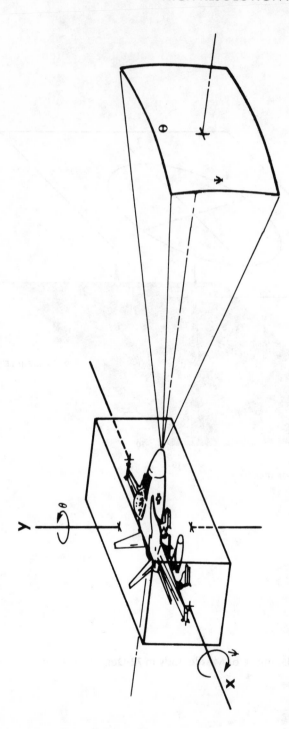

Figure 3-46. Two-Dimensional Synthetic Aperture Formed by Object Rotation.

dimensional case, experiments to demonstrate the feasibility of this process have not been attempted and constitute a potential future activity.

REFERENCES

1. Rihaczek, A. W., *Principles of High Resolution Radar,* New York, McGraw-Hill, Company, 1969, pp. 51-56.
2. Skolnik, M. I., *Radar Handbook,* New York: McGraw-Hill, Co., 1970, pp. 20.2-20.4.
3. Harris, F.J., "On the Use of Windows for Harmonic Analyses with the Discrete Fourier Transform," Proc. IEEE, Vol. 66, No. 1, pp. 51-83, January 1978.
4. Mensa, D.L., "A Linear FM System for High Resolution Radar Backscattering Measurements" Pacific Missile Test Center Report, TM-80-24, Point Mugu, California, August 1980.
5. Brown, W. M. and L.J. Porcello, "An Introduction to Synthetic Aperture Radar," IEEE Spectrum, pp. 52-62, September 1969.
6. Harger, R. O., *Synthetic Aperture Radar Systems, Theory, and Design,* New York, Academic Press, 1970.
7. Tomiyasu, K., "Tutorial Review of Synthetic-Aperture Radar (SAR) with Applications to Imaging of the Ocean Surface," IEEE, Vol. 66, pp. 563-583, May 1978.
8. Rihaczek, A. W., *Principles of High Resolution Radar,* New York, McGraw-Hill Company, 1969, p. 451.
9. Otnes, R. K. and L. I. Enochson, *Digital Time Series Analysis,* New York, John Wiley and Sons, Inc., 1972, pp. 420-424.

Focused Synthetic
Aperture Processing

In the preceding chapter a method for two-dimensional imaging of rotated objects was described. The resolution in range was shown to be determined by the signal bandwidth and the resolution in cross-range by the angular extent of the synthetic aperture. Signal processing consisted almost exclusively of Fourier transforms, which can be implemented by highly efficient FFT algorithms. The Fourier transform is the exact operation required for range processing and no approximations are involved in its application. However, its use in synthetic aperture processing constitutes an approximation which limits resolution. The image produced is subject to focusing errors which are predominantly range-dependent; consequently, Fourier transforming of signals which are received from a rotating object results in an imaging system which is space-variant and does not provide a diffraction-limited image throughout the object space. In this chapter, methods are considered for focusing the synthetic aperture in order to obtain diffraction-limited, space-invariant, point-spread functions. As shown in the preceding chapter, focusing errors are phase factors which depend on the wavelength of the irradiation. In the following section we develop the focusing process for the case of CW irradiation with a single wavelength. In the subsequent section, the analysis is extended to include polychromatic or wide-band irradiation.

Based on "Aperture Synthesis by Object Rotation in Coherent Imaging," IEEE Transactions on Nuclear Science, Vol. NS-27, No. 2, pp. 989-998, April 1980. © 1980 IEEE.

Synthetic Aperture Focusing for CW Signals

Consider the problem of obtaining a two-dimensional image of an object by processing reflected signals using a source and receiver located in the cross-sectional plane. The cross section is characterized by a two-dimensional distribution of scattering centers, denoted by g(x, y), which is termed the reflectivity density function. The geometry is shown in figure 4-1 which depicts an object being uniformly illuminated by a stationary CW source while reflected signals are recorded as a function of the object rotation angle*. The observation distance is assumed to be sufficiently greater than the object extent so that iso-range contours can be assumed to be straight lines normal to the line-of-sight. The objective is to obtain a reconstruction of g(x, y) by processing the received signals from various aspect angles. The term "signal" is taken to mean the complex envelope (phase and amplitude) of the received CW carrier.

The formulation is directly applicable to planar objects and to three-dimensional objects when a fan beam is used to illuminate selectively a particular cross section of the object. If a three-dimensional object is uniformly illuminated, g(x, y) is interpreted as the projection of the three-dimensional reflectivity density onto the xy-plane. The imaging process is tomographic in nature because the reconstructed image can be a slice of the object obtained from measurements of signals propagating in the plane of the slice. This is the case when a fan beam is used. The process, however, differs from conventional x-ray tomography (1)-(4) in which the image is reconstructed using measurements of line integrals or projections of the object density. As shown in the following analysis, the imaging process being described reconstructs the image from measurements of total reflected signal.

Assume the observations are started with the sensor aligned with the y axis; x is then the cross-range coordinate and y is the range coordinate. When the object is rotated through any angle θ, the range and cross-range coordinates become v and u, respectively. For a fixed θ, the signal from a particular range, v, is given by the reflectivity density integrated over u for a fixed v. This assumes that all reflectors from a given range add in-phase or, equivalently, that the reflection coefficients are of equal phase.

$$p(v;\theta) \propto \int_{-\infty}^{+\infty} g_\theta(u,v)\, du \qquad (4\text{-}1)$$

---

*When imaging complicated reflective objects, parts of the object may be shadowed by other parts. In our development we assume that shadowing does not occur.

Chapter 4 105

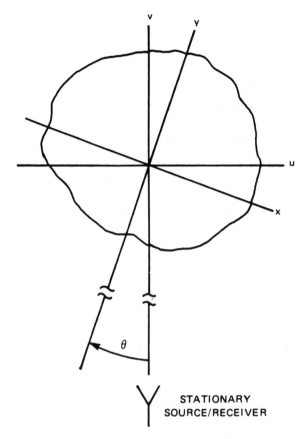

Figure 4-1. Imaging Geometry for a Planar Object.

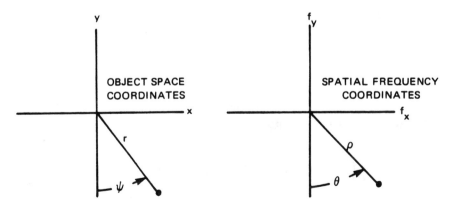

Figure 4-2. Space and Frequency Domain Coordinates.

where

$$g_\phi(u,v) = g(x,y)$$

$$x = u \cos \theta - v \sin \theta$$

$$y = u \sin \theta + v \cos \theta$$

the function $p(v;\theta)$ is a line integral consisting of a projection of the reflectivity density onto the v axis. It is a one-dimensional function of the single variable v; with the second variable, θ, interpreted as a parameter to denote the orientation of a projection.

The total signal received is the integral of signals along the projection modified by the phase factor determined by the round-trip phase. If the propagation velocity in the medium is constant, the signal is expressed within a generally complex constant by:

$$G(\theta) = \int_{-\infty}^{+\infty} p(v;\theta) \exp(-j4\pi v/\lambda)\,dv = \iint_{-\infty}^{+\infty} g_\theta(u,v) \exp(-j4\pi v/\lambda)\,du\,dv$$

$$(4\text{-}2)$$

The integrals can be expressed with infinite limits because the integrand is zero outside the bounded object.

The relation $v = y \cos \theta - x \sin\theta$ allows equation (4-2) to be expressed as:

$$G(\theta) = \iint_{-\infty}^{+\infty} g(x,y) \exp[-j(4\pi/\lambda)(y \cos \theta - x \sin \theta)]\,dx\,dy$$

$$(4\text{-}3)$$

If the variables f_x and f_y are defined as: $f_x = 2(\sin\theta)/\lambda$ $f_y = -2(\cos\theta)/\lambda$
equation (4-3) can be written as:

$$G(f_x, f_y) = \iint_{-\infty}^{+\infty} g(x,y) \exp[j2\pi(f_x x + f_y y)]\,dx\,dy$$

$$(4\text{-}4)$$

Equation (4,4) has been expressed as a two-dimensional Fourier transformation. If $G(f_x, f_y)$ were known for f_x, f_y running from $-\infty$ to $+\infty$, $g(x, y)$ could be readily determined by an inverse Fourier transformation resulting in:

$$g(x,y) = \int\int_{-\infty}^{+\infty} G(f_x, f_y) \exp\left[-j2\pi(f_x x + f_y y)\right] df_x df_y \qquad (4\text{-}5)$$

Equations (4-4) and (4-5) would then form a two-dimensional transform pair symbolized by:

$$g(x,y) \Leftrightarrow G(f_x, fy) \qquad (4\text{-}6)$$

$g(x, y)$ is the spatial function, $G(f_x, f_y)$ is the corresponding spectrum; f_x and f_y are planar wave spatial frequency components in the x and y directions, respectively.

Equation (4-5) shows that $g(x, y)$ could be determined by Fourier transforming $G(f_x, f_y)$ which is related to the observed data. Because $g(x, y)$ is space-limited, $G(f_x, f_y)$ extends over the entire $f_x f_y$-plane. However, the constraints $f_x = 2(\sin\theta)/\lambda$ and $f_y = -2(\cos\theta)/\lambda$ show that the actual data are limited to values of $G(f_x, f_y)$ falling along a circle in the $f_x f_y$-plane with radius $2/\lambda$. In the following analysis, we determine how the limited sample of the spatial spectrum afforded by the measured values $G(\theta)$ affects the reconstructed image.

Let $g(x, y)$ be the reconstructed reflectivity density obtained by setting $G(f_x, f_y) = 0$ everywhere except on a circle of radius $2/\lambda$. The radial symmetry allows simplification by converting to polar coordinates, defined in figure 4-2. Now, $g_R(x,y)$ and $g_P(r,\psi)$ are spatial descriptions of the object in rectangular and polar coordinates; $G_R(f_x,f_y)$ and $G_P(\rho,\theta)$ are the corresponding spectral functions expressed in rectangular and polar coordinates. The expression for the reconstruction in polar coordinates is:

$$\hat{g}_P(r,\psi) = \int_0^\infty \int_0^{2\pi} G_P(\rho,\theta)\, \delta(\rho - 2/\lambda) \exp\left[-j2\pi\rho r \cos(\theta - \psi)\right] d\theta\, \rho\, d\rho$$

$$= 2/\lambda \int_0^{2\pi} G_P(\theta) \exp\left[-j4\pi r \lambda^{-1} \cos(\theta-\psi)\right] d\theta \qquad (4\text{-}7)$$

The values $G_P(f_x, f_y)$ are precisely the observed data $G(\theta)$ as described in equation (4-3).

Equation (4-7) expresses the response of a circular line aperture which surrounds the object and is focused to each point in the object space. The exponential term is the phase correction required to bring signals received from an object point at r, ψ to an in-phase condition. Although equation (4-7) is expressed as a single integral as a result of the radial symmetry, it constitutes a two-dimensional Fourier transform. The two-dimensional operation is required to focus the aperture to all points in the object plane.

The point-spread function of the imaging system can be obtained by employing the transform relation:

$$\widehat{g_P}(r, \psi) = \mathcal{F}\left\{[G_P(\rho,\theta)]\,[\delta(\rho - 2/\lambda)]\right\} \qquad (4\text{-}8)$$

Because $\widehat{g_P}(r, \psi) = \mathcal{F}[(G_P(\rho,\theta)]$,   equation (4-8) becomes:

$$\widehat{g_P}(r, \psi) = g_P(r, \psi) \overset{\star}{_2} h(r, \psi) \qquad (4\text{-}9)$$

where $h(r, \psi) = \mathcal{F}[\delta(\rho - 2/\lambda)]$

and the symbol $\overset{\star}{_2}$ denotes two-dimensional convolution.

The calculation of $h(r, \psi)$ is performed by a Fourier integral in polar coordinates.

$$h(r, \psi) = \int_0^\infty \int_0^{2\pi} \delta(\rho - 2/\lambda)\, \exp[-j2\pi\rho r \cos(\theta - \psi)]\, d\theta\, \rho\, d\rho$$

$$h(r) = (2/\lambda) \int_0^{2\pi} \exp[-j2\pi r(2/\lambda) \cos(\theta - \psi)]\, d\theta = (4\pi/\lambda) J_0(4\pi r/\lambda)$$

$$(4\text{-}10)$$

The reconstruction $\widehat{g}(r, \psi)$ is a convolution of the object reflectivity distribution, $g(r, \psi)$, with the function $h(r,\psi)$, which is the point spread function of the imaging process. The point-spread function, expressed by equation (4-10), exhibits radial symmetry with a central lobe width (radial distance between zeroes adjacent to the peak) of 0.4λ; the zero-crossing space is not precisely constant with r and the first side lobe is 8 decibels (dB) below the peak. Its magnitude squared is plotted in figure 4-3.

The interpretation of the point-spread function of the process is that object points will be imaged as a central peak of width 0.4λ, surrounded by monotonically decreasing circular sidelobe ridges and valleys. The imaging process is linear and shift invariant; therefore, the form of the response is uniform throughout the object space. The image response to multiple point objects is the superposition of point-spread functions, each weighted by the amplitude of the object point and shifted to the spatial coordinates of the object point. Because the first zero of the point-spread function occurs at a radial distance of 0.2λ from the peak, two point-objects of comparable amplitude separated by more than 0.2λ will produce an image in which they are resolved.

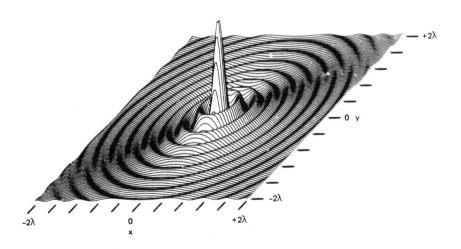

Figure 4-3. Intensity of the Point-Spread Function for the Imaging Process.

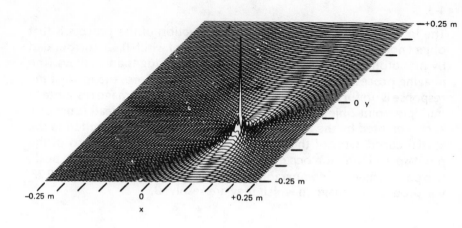

Figure 4-4. Intensity of Reconstructed Image of a Line Object (λ=0.03 m).

Figure 4-4 shows the intensity of the image reconstructed from data computed to simulate a signal received from a point object. The simulated wavelength was 0.03 meter and the image field was calculated for an area of 0.5 x 0.5 meter and consisted of 128 x 128 pixels. The object point was placed at a radial distance of 0.15 meter and 300 uniform angular increments were used for a complete rotation. The reconstructed image conforms to the theoretical point-spread function and, not being centered, confirms the shift invariance of the imaging process. Small artifacts result from plotting the radially symmetric response on a rectangular format.

In order to confirm the practical performance of the imaging process, a microwave measurement of a dual point object was conducted and the image reconstructed from the recorded data. Synthetic data corresponding to parameters duplicating those of the test were also generated and processed. The test object was a pair of identical rods 0.64 cm in diameter and 34 cm in length. The rods, separated by 17 cm and mounted vertically on a rotating support, were irradiated by a radar with a wavelength of 0.06 m from a distance of approximately 20 m. The rotation axis was parallel to the rods and normal to the radar line-of-sight. The test object is essentially a pair of line objects because the rod diameter is small relative to the wavelength; the projection of the line objects onto the xy-plane is a pair of points. The reconstructed image field is a 128 x 128 cell array spanning 0.5 x 0.5 m. Figures 4-5 and 4-6, displaying the intensity of the images reconstructed from real and synthetic data, show excellent correlation. The image artifacts are induced by interference between the periodic sidelobe structures of the two object responses.

The imaging process can be considered from a spatial spectrum viewpoint with the reconstructed image interpreted as a synthesis of elemental plane-wave components. Consider the reconstruction of a single point-object of unit amplitude with coordinates (x_o, y_o).

$$G_R(f_x, f_y) = \int\int_{-\infty}^{+\infty} g_R(x, y) \exp\left[j2\pi(f_x x + f_y y)\right] dx\, dy \qquad (4\text{-}11)$$

If $g_R(x, y) = \delta(x - x_o, y - y_o)$, equation 4-11 becomes:

$$G_R(f_x, f_y) = \exp\left[j2\pi(f_x x_o + f_y y_o)\right] \qquad (4\text{-}12)$$

The spatial spectrum of equation (4-12) has constant magnitude and linear phase; that is, the spectrum contains all spatial frequencies and has a phase linear with frequency and proportional to the displacement of the object. The reconstruction of the spatial function is obtained by evaluating equation (4-7) for $G(\theta)$ corresponding to $g_R(x,y) = \delta(x - x_o, y - y_o)$. Algebraic manipulations after converting the result to rectangular coordinates yield:

$$\hat{g}_R(x, y) = (2/\lambda) \int_0^{2\pi} \exp\left\{-j2\pi[(2/\lambda) \sin\theta(x - x_o)\right.$$
$$\left. - (2/\lambda) \cos\theta(y - y_o)]\right\} d\theta \qquad (4\text{-}13)$$

Equation (4-13) indicates that the reconstruction is formed by a spectrum of planar wave components each with spatial frequency of $2/\lambda$, directed over the complete range of angles from 0 to 2π. The reconstruction can be visualized as a superposition of sinusoidally corrugated infinite sheets of fixed corrugation spacing $\lambda/2$ and orientations varying from 0 to 2π.

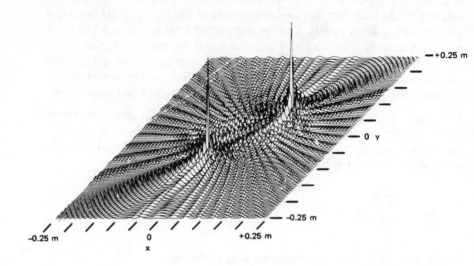

Figure 4-5. Intensity of Image Reconstructed From Measured Data
for Two-Line Objects (λ=0.06 m).

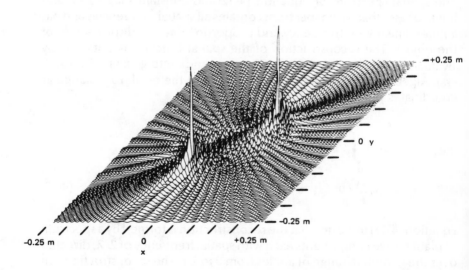

Figure 4-6. Intensity of the Image Reconstructed From Synthesized
Data for Two-Line Objects (λ=0.06 m).

The amplitude of the corrugation is defined by $|G(\theta)|$; the displacement normal to the corrugations is defined by the phase Arg $[G(\theta)]$, and the orientation is $\theta = \tan^{-1} (f_x/-f_y)$. Figure 4-7 shows the real part of the elemental function of the reconstruction process. The superposition of such elemental functions represented in equation (4-13) yields a central peak whose width is less than one-half the corrugation spacing, $\lambda/4$, with the peak occuring at the point where the corrugations add in phase.

Point-spread functions corresponding to partial circular apertures of 180, 90, 60, 30, and 10 degrees are shown in figures 4-8 through 4-12. The plots were generated by carrying out a numerical integration of equation (4-10) over a limited range of θ. The range axis (y) is normal to the center of the aperture. The point-spread function broadens most rapidly in the range direction as the aperture angle decreases. As the angular extent of the aperture decreases, the widths of the point-spread function in the range and cross-range directions approach the limits derived in the preceding chapter:

$$\Delta x = \frac{\lambda}{2(\Delta\theta)} \; ; \; \Delta y = \frac{\lambda}{(\Delta\theta)^2} \tag{4-14}$$

As the angular extent of the aperture increases to surround the object, the point-spread function becomes radially symmetric with width 0.4λ.

The operations required to implement the imaging process consist of obtaining received data samples corresponding to uniformly spaced positions on a circular ring in the spatial frequency domain and performing a two-dimensional Fourier transform. This operation inherently focuses the aperture throughout the object field and produces a point-spread function which is diffraction-limited and shift invariant.

The reconstructed image is a good approximation to the reflectivity density of the objects considered in view of the very limited sampling of the spatial spectrum. The relatively large sidelobes in the point-spread function result from the discontinuous nature of the annulus in the spectral domain. The reconstruction obtained by setting all unmeasured values to zero is subject to question. Because the object is space-limited, the spatial spectrum extends to infinite frequencies and therefore cannot be zero everywhere outside the annular sample. Thus, it can be stated unequivocally that the unmeasured values of the spectrum cannot be everywhere zero. The possibility of improving the reconstruction by extrapolating the spectral data is considered in the next chapter.

Although the imaging process can be analyzed from a spatial view-point, additional insight is obtained by considering the object from a spatial frequency viewpoint. Because the object image and its spatial spectrum are related by a Fourier transform, a description of the spectrum over the entire frequency plane is equivalent to complete knowledge of the object in the spatial domain. Such knowledge, there-fore, would provide an exact reconstruction. The data obtained from object rotation using a single wavelength constitute samples of the spectrum along a circular arc of radius $2/\lambda$ where the angular extent of the arc is equal to the angular rotation of the object. Figure 4-13 illus-trates the spatial frequency domain representation. For small angular apertures, the circular arc is suitably approximated by a straight-line segment and the spatial response is adequately approximated by a one-dimensional Fourier transform of the observed data. From this consider-ation, the nature of the approximation incurred by using a one-dimensional Fourier transform for synthetic aperture processing can be visualized. Unfocused synthetic aperture processing is equivalent to approximating a circular segment of the spatial spectrum as a linear segment. The processing required to focus the aperture is the evaluation of the two-dimensional Fourier transform along a circular arc.

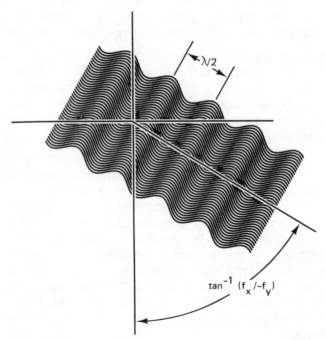

Figure 4-7. Elemental Function of Reconstruction Process.

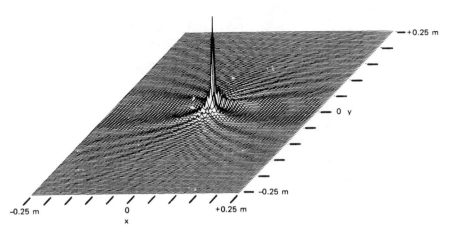

Figure 4-8. Point-Spread Function for 180-Degree Partial Aperture (λ=0.03 m).

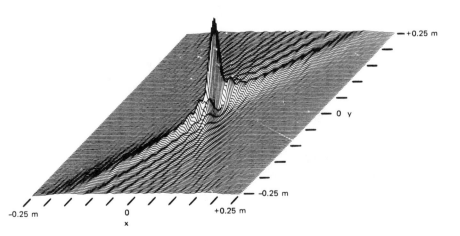

Figure 4-9. Point-Spread Function for 90-Degree Partial Aperture (λ=0.03 m).

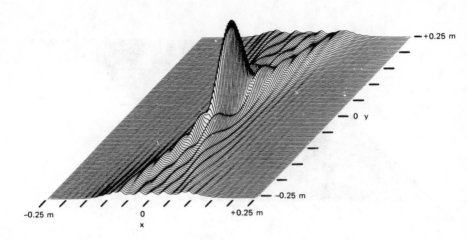

Figure 4-10. Point-Spread Function for 60-Degree Partial Aperture (λ=0.03 m).

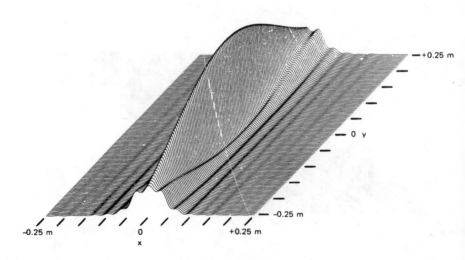

Figure 4-11. Point-Spread Function for 30-Degree Partial Aperture (λ=0.03 m).

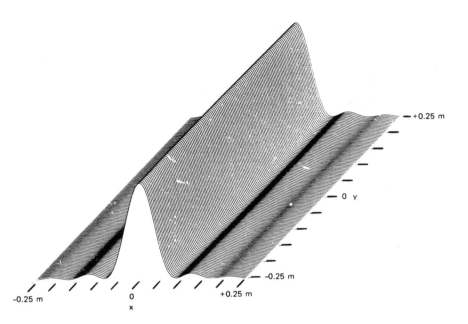

Figure 4-12. Point-Spread Function for 10-Degree Partial Aperture (λ=0.03 m).

The principal advantage of focused synthetic-aperture processing using CW signals is the high degree of two-dimensional spatial resolution achievable without use of wideband signals. In contrast, the conventional method of obtaining range resolution by time-delay sorting requires wide-band signals. The disadvantages of the CW method are the high sidelobes in the point-spread function which limit the dynamic range of the image and the degradation in image resolution if object points cannot be viewed over large angles ($\geq\pi$ radians). The method, therefore, allows high resolution imaging of sparse arrays of object points which are small compared to a wavelength and are of nearly equal magnitude.

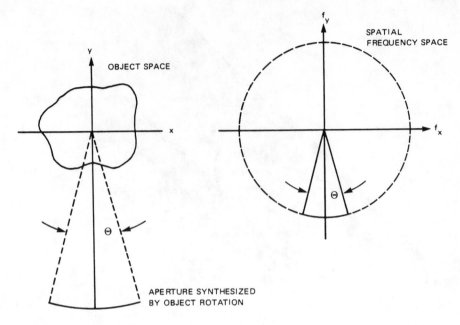

Figure 4-13. Spatial Frequency Sampling Obtained From Object Rotation.

Synthetic Aperture Focusing for Wideband Signals

The quality of images reconstructed from a focused synthetic aperture using CW signals is limited by the restricted sampling of the spatial spectrum along a ring. In order to improve the reconstruction, the spatial spectrum must be determined over a greater part of the frequency plane. The use of wide-band signals discussed in the preceding chapter affords the possibility of extending the region in which the spectrum can be measured. Because the use of a CW signal provides a measurement of the spatial spectrum on a ring of radius $2f/c$, using a signal bandwidth of Δf extends the measurable spectral region to an annulus with radial extent $2\Delta f/c$.

We develop the concept of coherently processing a wide-band signal to form a focused synthetic aperture in order to achieve diffraction-limited imaging. Consider a planar object irradiated by a monochromatic signal of frequency f from a distance R_0 and angle θ, as shown in figure 4-14. The variables r and ψ locate a point in the coordinate plane fixed to the object. Assuming $R_0 \gg r$ establishes the iso-range lines as parallel and normal to the line-of sight. The signal reflected from an object point is expressed by:

$$v_R(t) = R_e \left\{ \exp(j2\pi ft) \, \sigma \exp \left[-j4\pi f c^{-1}[R_0 - r\cos(\psi - \theta)] \right] \right\} \qquad (4\text{-}15)$$

The first bracketed quantity expresses the temporal variations due to the transmitted signal carrier; the second bracketed quantity is the complex envelope which represents the modulation imposed on the carrier by the propagation delay and object properties; δ is a complex constant which includes the reflectivity of the object and the attenuation due to the propagation distance. The term $\exp(j4\pi f R_0/c)$ in the complex envelope expression is a constant phase factor which plays no significant role; for convenience it can be included in δ. The remaining term determines the phase variations in the received signal; it is the significant part of the complex envelope and is given by:

$$G(f,\theta) = \exp \left[j4\pi \, f r c^{-1} \cos(\theta - \psi) \right] \qquad (4\text{-}16)$$

The general behavior of the complex envelope can be shown as a function of f and θ by displaying the iso-phase lines which satisfy the equation:

$$2\pi n = 4\pi \, f r c^{-1} \cos(\theta - \psi) \qquad (4\text{-}17)$$

Integer values of n map loci of points in the f-θ plane for which the phase is uniform and equal to $+2\pi n$. Figure 4-15 shows plots of equation (4-17) for r = 1.5 m, n ranging from 0 to 100, f from 0 to 10 GHz, and $(\theta - \psi)$ from $-\pi/2$ to $+\pi/2$. The iso-phase lines indicate the cyclic variations of the complex signal envelope as a function of f and θ. Examination of figure 4-15 indicates that the local spacing of the iso-phase lines in the vertical (f) and horizontal (θ) directions are inversely proportional to the range and cross-range coordinates of the object, respectively. This can be verified by differentiating equation (4-17):

$$\left. \frac{dn}{df} \right|_{\theta \, = \, constant} = 2c^{-1} \left[r\cos(\theta - \psi) \right] \qquad (4\text{-}18)$$

$$\left. \frac{dn}{d\theta} \right|_{f = constant} = 2fc^{-1} \left[r \sin (\theta - \psi) \right] \qquad (4\text{-}19)$$

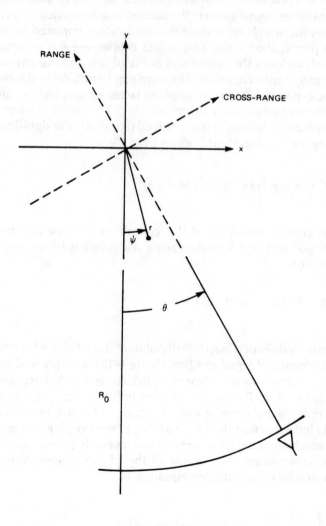

Figure 4-14. Planar Object Imaging Geometry.

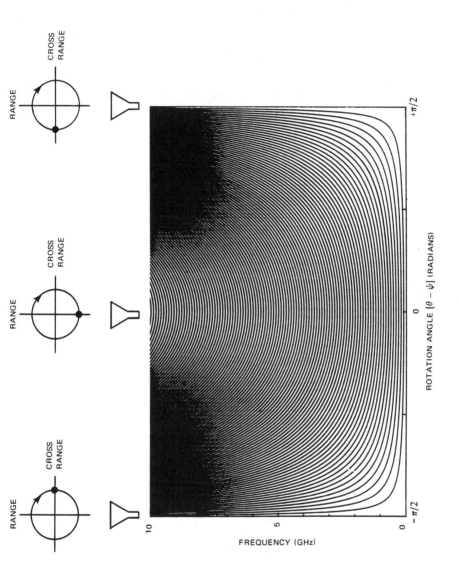

Figure 4-15. Iso-Phase Contours of Complex Envelope
for Rotating Point-Object.

By referring to figure 4-14, the bracketed quantities in equations (4-18) and (4-19) are seen to be the coordinates of the reflecting point measured along, and normal to the line-of-sight; these distances are the range and cross-range coordinates of the object point relative to the sensor. For a fixed angle, the spacing of the iso-phase lines is uniform and varies inversely with range; for a fixed temporal frequency the spacing of the iso-phase lines varies inversely with cross-range.

Over limited regions of the f-θ plane, the iso-phase lines are approximately linear and uniformly spaced; over such regions, therefore, the variations in the complex envelope are uniquely described by the horizontally- and vertically-directed spatial frequency components of the iso-phase line plot. A two-dimensional Fourier transform of the complex signal over the limited region will produce an image of the point properly located in the range, cross-range space. For an object consisting of an array of point reflectors, the f-θ map of the complex envelope is a superposition of maps for single reflectors, properly shifted in position and amplitude in accordance with location and magnitude of the point reflectors. The superposition properties of the Fourier transform provide the decomposition required to image individual reflectors. If the processing is limited to a restricted region, the two-dimensional imaging is represented by a scaled Fourier-transform pair between (x, y) and (f, θ), as illustrated symbolically in figure 4-16.

The method described has been applied to process microwave data optically (5), but imaging can be performed by processing the sampled received signal using a computer (6). If optical processing is used, the recording of the fringe pattern represented by figure 4-15 is by necessity real and the Fourier transform will exhibit conjugate images and background illumination due to undiffracted light. If the processing is performed digitally on a record of the complex envelope, an unambiguous image is obtained.

The variations of the complex envelope in the vertical direction on figure 4-15 represent the object response as a function of the temporal frequency of the transmitted signal. This is precisely the response produced by the linear FM system described in the preceding chapter with the following minor modification required to account for the fixed range R_0. If the reference signal of the linear FM system is subjected to a delay equivalent to the range R_0, the resulting signal can be expressed by:

$$v(t) = R_e \left\{ E_0 \Pi(t/T) \exp \left[-j4\pi(BtR/cT + f_0 R/c) \right] \right\} \qquad (4\text{-}20)$$

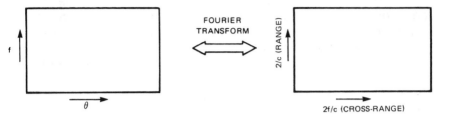

Figure 4-16. Fourier Transform Relation Between Complex Envelope
and Object Point Coordinates.

where B = total swept bandwidth
    R = differential range measured along line-of sight (range - R_0)
    f = center frequency of swept bandwidth
    T = chirp period

Substituting f = Bt/T for the instantaneous frequency deviation of the transmitted signal yields:

$$v(f) = R_e \left\{ E_0 \Pi(f/B) \exp \left[-j4\pi(fR/c + f_0 R/c) \right] \right\} \qquad (4\text{-}21)$$

Because the instantaneous frequency of the transmitted signal varies linearly during the chirp period, equation (4-21), suitably scaled, represents the variation of the complex envelope as a function of the instantaneous frequency of the irradiating signal. With the exception of the constant phase term represented by the last term of the exponent, equation (4-21) is a real signal representation of the complex envelope of equation (4-16). The output of the coherently demodulated FM system can be made to modulate the intensity of a vertical line on a cathode ray tube or a film transparency; recording an ensemble of closely spaced vertical lines, each corresponding to a unique observation angle, results in a fringe pattern such as that represented by figure 4-15. The recording, reminiscent of a hologram, contains information on the magnitude and spatial position of the object point, which can be coherently processed by optical means.

Whether optical processing is used on the real record of intensity or computer processing on the complex signal envelope, the image resolution is determined by the size of the f - θ region used. When processing consists of a Fourier transform, the usable region is limited to a maximum size such that the iso-phase lines are essentially linear. Fourier transforming data from a larger region in order to increase resolution results in image degradations as described in the preceding chapters. An optical approach for improving the achievable resolution, termed sectional processing, is presented in (7). By this method, larger processing apertures are obtained using lenses as matched filters to compensate for curvature of the fringes in the f - θ plane. The limitation of this method is that a different matched filter is required to optimize the image of each part of the object; the final image must then be produced by photographically superimposing sections which have been individually processed.

The effect of phase curvature being described is identical to that discussed in preceding chapters dealing with Doppler imaging and synthetic aperture processing; the cause is the nonlinear variation in the phase of received signals resulting from the circular motion of rotating objects. In the section to follow, we develop methods of compensating for phase curvature by using *a priori* knowledege that the motion is circular.

We note from equation (4-17) that the lines of constant phase correspond to a constant product f cos (θ - ψ), which, by expanding in a trigonometric identity can be written as:

$$f \cos (\theta - \psi) = f \sin \theta \sin \psi + f \cos \theta \cos \psi \qquad (4\text{-}22)$$

Equation (4-22) represents a constant expressed in terms of a set of mutually orthogonal coordinates, f sin θ and f cos θ, which have been rotated through an angle ψ. It is thus possible to determine a transformation which maps the curved lines of figure 4-15 into a set of straight parallel lines. The conformal mapping of G(f, θ) into a new function, G'(X,Y), is accomplished by the following relations, illustrated in figure 4-17:

$$X = f \sin \theta$$
$$Y = -f \cos \theta \qquad (4\text{-}23)$$

The function $G(f,\theta) = \exp[j4\pi rc^{-1} \cos(\theta - \psi)]$ of equation (4-16), therefore maps onto the function:

$$G'(X,Y) = \exp[j4\pi rc^{-1}(X \sin \psi - Y \cos \psi)] \qquad (4\text{-}24)$$

Substituting the rectangular coordinates of the point object shown in figure 4-14 into equation (4-24) yields:

$$G'(X,Y) = \exp[j4\pi c^{-1}(Xx + Yy)] \qquad (4\text{-}25)$$

If the iso-phase lines of the complex envelope expressed in equations (4-24) or (4-25) are plotted in the X, Y space, they will consist of a set of parallel lines rotated from the X axis by the angle of ψ, as shown in figure 4-18. The following observations about the fringe pattern plotted in the X-Y space can be made.

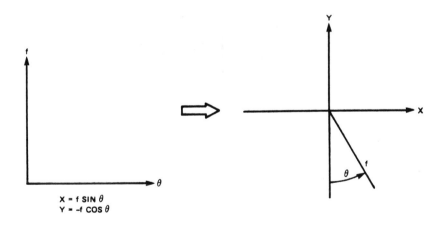

Figure 4-17. Transformation for Mapping $G(f,\theta)$ Into $G'(X,Y)$.

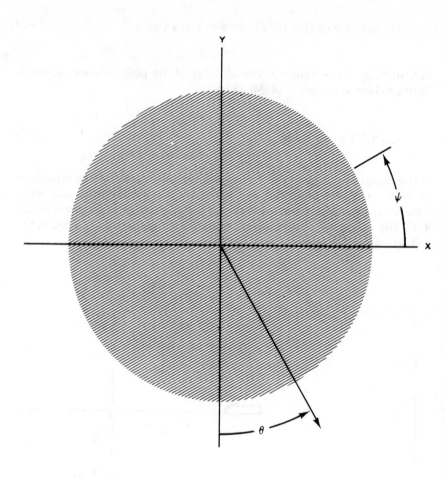

Figure 4-18. Iso-Phase Contours of Complex Envelope for Rotating Point Object in Transformed Coordinates.

1. The fringes for a point object are parallel lines with orientation determined by the angular position, ψ, of the point object.

2. The fringe spacing is inversely proportional to the radial coordinate, r, of the object and is given by c/2r.

3. The fringe pattern for a point located on the y axis is a set of horizontal lines; that for a point on the x axis is a set of vertical lines. A point at the origin (r = 0) produces no fringes because the received signal phase is constant.

4. The radial limit of the fringe pattern is determined by the maximum temporal frequency of the irradiating signal; if the irradiating signal frequency is limited to a range $f_0 \pm \Delta f/2$, the fringe pattern will be an annular section of radial extent equal to Δf.

The transformation of equation (4-23) precisely corrects the phase curvature and therefore constitutes the required focusing correction. Subsequent to this correction, a two-dimensional Fourier transform can be applied to all the available data. By this method, spatial frequencies can be separated to produce images of object points properly located in a two-dimensional space. The complex envelope recorded in a polar format represents a superposition of planar waves with magnitudes and orientations related to the amplitudes and locations of object points. The realization that the polar transformation provides the necessary correction for focusing is attributed to Walker (8) who applied the technique to optically record data directly in a polar format for subsequent Fourier transformation.

The Fourier transform relation between G'(X,Y) and the spatial distribution of reflectivity g(x, y) can be formalized as follows. For an ensemble of point objects, denoted by g(x, y), the complex envelope is the superposition of terms as equation (4-24) weighted by g(x, y) resulting in:

$$G'(X,Y) = \iint_{-\infty}^{+\infty} g(x,y) \exp\left[j2\pi(2c^{-1}Xx + 2c^{-1}Yy)\right] \, dx \, dy \qquad (4\text{-}26)$$

Equation (4-26) is a two-dimensional Fourier transform between the pair $(x,y) \Longleftrightarrow (2X/c, 2Y/c)$. The spatial function, g(x, y) is obtained from the Fourier transform of (4-26) which is:

$$\hat{g}(x,y) = \iint_{-\infty}^{+\infty} G'(X,Y) \exp\left[-j2\pi(2c^{-1}Xx + 2c^{-1}Yy)\right] \, d(2X/c) \, d(2Y/c) \qquad (4\text{-}27)$$

This formally establishes the Fourier transform relation between the complex envelope function mapped in the X-Y space and the spatial distribution of the object's reflectivity. When a single temporal frequency, f_0, is used in a CW measurement, $G'(X, Y)$ is measured only on a circular ring of radius $2f_0/c$ as shown in figure 4-19. In this case, $G'(X, Y) = \delta[2c^{-1}(f - f_0)]$ where $f^2 = X^2 + Y^2$. We note that this is precisely the result of equation (4-7) derived in the preceding section. The result of the latter treatment is more general because it allows for wide-band irradiating signals; the results are, of course, applicable to the special case of CW irradiation.

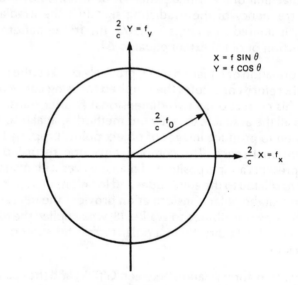

Figure 4-19. Spatial Spectrum Region Sampled
by a CW Measurement at Frequency f_0.

The preceding treatment demonstrates that mapping the complex envelope of the received signal in a polar format corrects the phase nonlinearities introduced by the object rotation and simultaneously focuses the entire aperture for all wavelengths. By subsequently processing the corrected data with a two-dimensional Fourier transform, a diffraction-limited image is obtained. In order to reconstruct perfectly the object reflectivity density, the spatial spectrum must be known over the entire spatial frequency plane; this corresponds to an impulse point-spread function. In all practical cases, however, the spatial spectrum is measurable only over a limited region and the point-spread function is spatially extended. The image reconstructed from a limited region of the spectrum is:

$$g(x,y) = \int\!\!\!\int_{-\infty}^{+\infty} H(X,Y)\, G'(X,Y)\, \exp\left\{-j2\pi[2c^{-1}Xx + 2c^{-1}Yy]\right\} d(2X/c)\, d(2Y/c)$$

$$= \int\!\!\!\int_{-\infty}^{+\infty} H(cf_x/2,\, cf_y/2)\, G(cf_x/2,\, cf_y/2)\exp[-j2\pi(f_x x + f_y y)]\, df_x\, df_y$$

$$(4\text{-}28)$$

where $H(X, Y)$ is a window function which accounts for the spectral bound. The point-spread function, $h(x, y)$, of the imaging process is determined by the inverse Fourier transform of the window function. Several radially symmetric functions and corresponding point-spread functions are shown in figures 4-20 through 4-24. These were determined analytically using two-dimensional transform pairs tabulated in (9). A notable feature of these plots is the nearly identical width of the central portion of all the point-spread functions. Evidently, the ability to discriminate between two closely-spaced equal point-objects is determined principally by the high spatial-frequency components. As the radial extent of the spectral window decreases, however, the sidelobes of the point-spread functions increase considerably. The sidelobes constitute image ambiguities which are not fully removed when a limited region of the spectrum is used for reconstruction. A heuristic interpretation of the process for reconstructing the image of a unit-amplitude point object located at $x, y = 0$ can be made as follows. Each point (f_x, f_y) in the spatial spectrum represents a complex planar wave component $\exp[-j4\pi\lambda^{-1}(f_x x + f_y y)]$. Each pair of diametrically opposed points, therefore, represents a cosinusoidally corrugated sheet with orientation corresponding to that of the spectral point-pair, as shown in figure 4-25. The spatial reconstruction corresponding to an extended spectrum, therefore, consists of the superposition of such corrugated sheets. A spectrum consisting of a circular ring, as shown in figure 4-20, represents a continuum of sheets of equal corrugation spacing and differing orientations. Upon superposition, these corrugated sheets have a common maximum; for all other spatial locations, some cancellation occurs. The incompleteness of this cancellation gives rise to residual sidelobes. As other spectral components are included in the superposition, sheets with different corrugation spacings are included and the cancellation is more complete. Finally, if all spectral components are employed, the complete cancellation results in a reconstruction which is maximum at the origin and zero elsewhere.

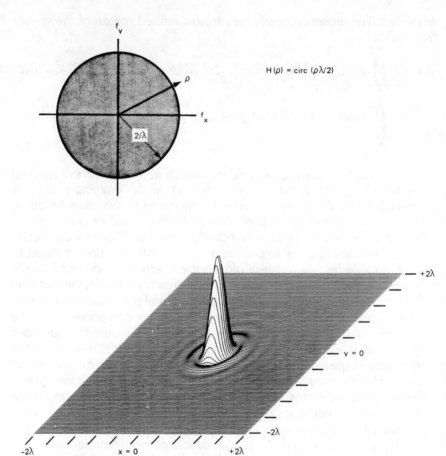

Figure 4-20. Intensity of Point-Spread Function
Corresponding to a Disk Spectral Window.

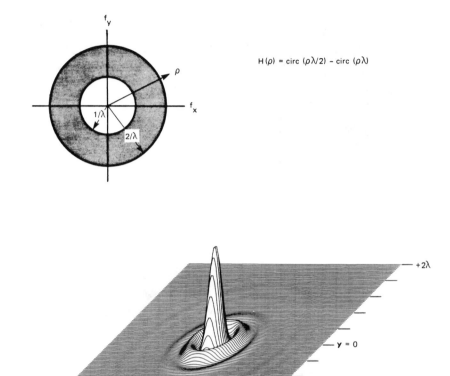

$$H(\rho) = \text{circ } (\rho\lambda/2) - \text{circ } (\rho\lambda)$$

Figure 4-21. Intensity of Point-Spread Function
Corresponding to a Wide Annulus Spectral Window.

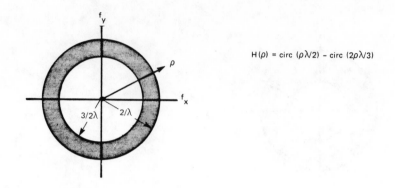

$$H(\rho) = \text{circ}\,(\rho\lambda/2) - \text{circ}\,(2\rho\lambda/3)$$

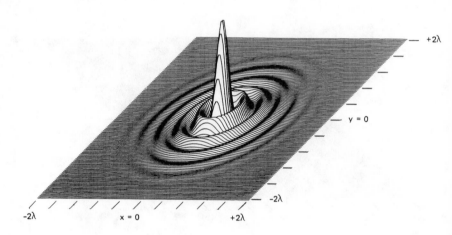

Figure 4-22. Intensity of Point-Spread Function
Corresponding to a Narrow Annulus Spectral Window.

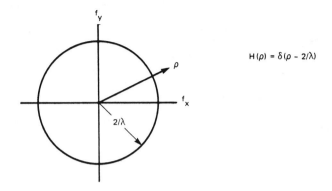

$$H(\rho) = \delta(\rho - 2/\lambda)$$

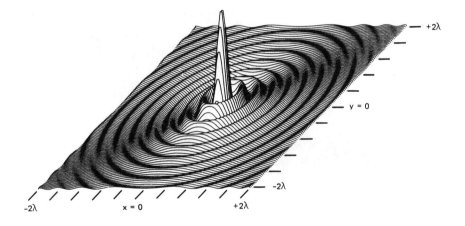

Figure 4-23. Intensity of Point-Spread Function
Corresponding to a Ring Spectral Window.

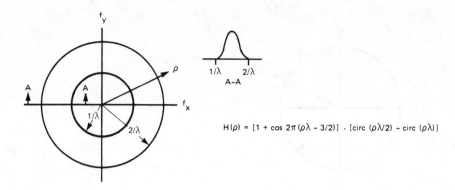

$$H(\rho) = [1 + \cos 2\pi (\rho\lambda - 3/2)] \cdot [\mathrm{circ}\ (\rho\lambda/2) - \mathrm{circ}\ (\rho\lambda)]$$

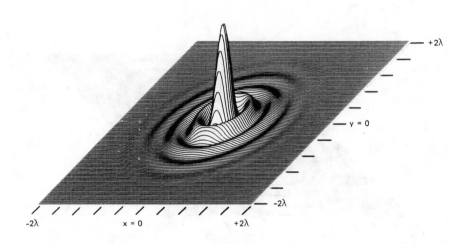

Figure 4-24. Intensity of Point-Spread Function
Corresponding to Tapered Annulus Spectral Window.

The presence of large sidelobes in the point-spread function is significant because it limits the ability to distinguish images of objects with small magnitude in proximity to those with large magnitude. In an attempt to reduce sidelobes, we may consider using tapered windows, analogous to the use of time windows, to reduce spectral leakage associated with finite observation intervals (10). Figure 4-24 shows the point spread function resulting from a radially tapered window which is an annulus with cosinusoidal cross section. The result, observed by comparing figures 4-24 and 4-21, is that the sidelobes of the point-spread function for the tapered window are increased over those for a uniform window of equal radial extent. A more complete analysis of the use of tapered two-dimensional windows applied to annular spectra for smoothing the sidelobes in the point-spread function is presented in the Appendix.

In some applications, measurements over a continuous range of temporal frequencies to achieve a spectral annulus of significant extent is not practical due to instrumentation limits. For such cases, we consider the alternative of performing a number of measurements using discrete temporal frequencies to sample the spatial spectrum over a number of concentric rings. In the limit of diminishing ring spacing, the continuous and discrete frequency cases become identical. Of practical interest is the imaging performance achievable with relatively few (<10) discrete frequencies. We expect that periodically sampling the spatial spectrum over discrete rings will result in image artifacts due to ambiguities associated with the discrete spectral sampling. Examples of analogous ambiguities in multi-frequency acoustic holography are shown in (11).

Owing to the linearity properties of the Fourier transform, the point-spread function and the image can be obtained by the superposition of responses corresponding to individual rings. Thus, if $2/\lambda_n$ is the radius of a discrete spectral ring, the point-spread function for N rings is:

$$h(r) = \sum_{n=1}^{N} J_0(4\pi r/\lambda_n) \tag{4-29}$$

Figures 4-26 and 4-27 show point-spread functions for window functions consisting of five and three spectral rings which are uniformly spaced between $1/\lambda$ and $2/\lambda$. The central part of the point spread functions are very similar to the corresponding continuous annulus of figure 4-21.

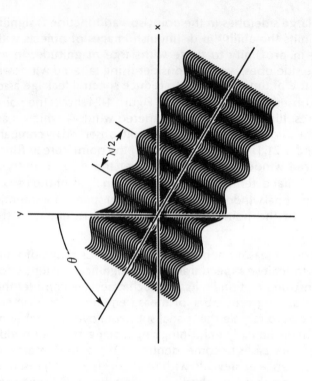

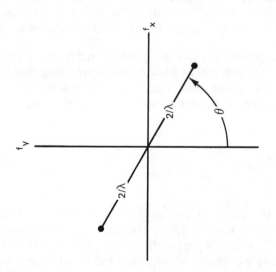

Figure 4-25. Spatial Function Corresponding to a Spectral Point-Pair.

Figure 4-27, the three-ring response, shows a circular artifact with radius 2λ which is due to periodic spectral sampling. The ambiguity artifact is not visible in figure 4-26 because it occurs outside the plotted range.

Figures 4-28 and 4-29 are expanded quadrants of the point-spread functions which illustrate the presence of periodic circular artifacts. A uniform spectral separation between rings, $\Delta\rho$, corresponds to a spatial period in the space domain, Δr, given by:

$$\Delta r = \frac{1}{\Delta\rho} \tag{4-30}$$

Substituting the coordinate relations shown in figure 4-19 leads to the equivalent expressions:

$$\Delta r = \frac{c}{2\Delta f} \tag{4-31}$$

$$\frac{\Delta r}{\lambda} = \frac{f}{2\Delta f} \tag{4-32}$$

where f indicates the temporal frequency of irradiation.

Using the above relations with $\Delta\rho = 0.5/\lambda$ and $0.25/\lambda$ for the three- and five-ring spectrum leads to $\Delta r = 2\lambda$ and 4λ, which coincide with the spacing between the ambiguity artifacts in figures 4-28 and 4-29. Note that the circular artifacts diminish in magnitude as their radius increases.

This means that close spacing of the spectral rings causes low-level artifacts at large distance from the central region of the point-spread function; of course, a small number of closely-spaced rings is equivalent to a narrow spectral annulus with attendant sidelobes.

The nature of the periodic ambiguities can also be established by the following development. The point-spread function for a single spectral ring of wavelength λ is $J_0(4\pi r/\lambda)$; the point-spread function for several rings is the superposition of $J_0(.)$ functions with differing arguments. In the superposition, the central peaks of the functions add constructively while the circularly symmetric sidelobes tend to cancel. At some unique radial distances, the sidelobes will reinforce and the superposition will

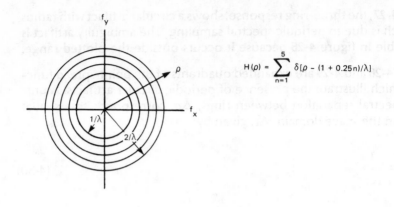

$$H(\rho) = \sum_{n=1}^{5} \delta[\rho - (1 + 0.25n)/\lambda]$$

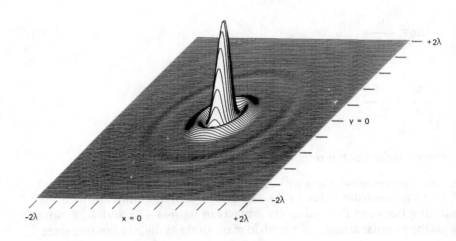

Figure 4-26. Intensity of Point-Spread Function Corresponding to a Five-Ring Spectral Window.

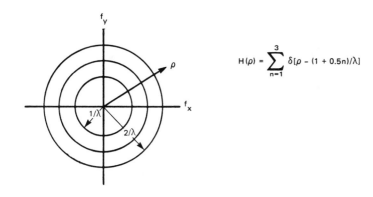

$$H(\rho) = \sum_{n=1}^{3} \delta[\rho - (1 + 0.5n)/\lambda]$$

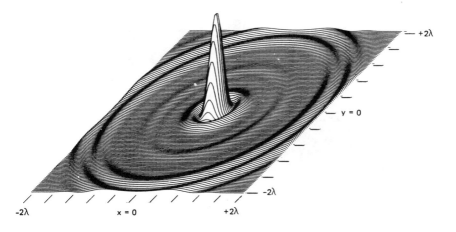

Figure 4-27. Intensity of Point-Spread Function Corresponding to a Three-Ring Spectral Window.

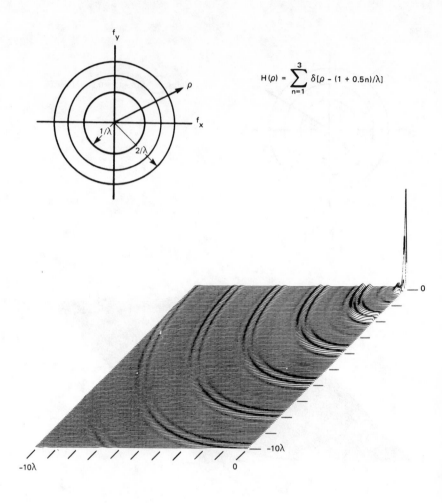

$$H(\rho) = \sum_{n=1}^{3} \delta[\rho - (1 + 0.5n)/\lambda]$$

Figure 4-28. Intensity of Expanded Point-Spread Function
for Three-Ring Spectral Window.

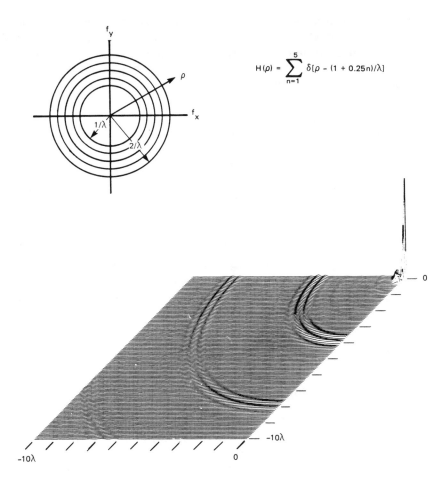

Figure 4-29. Intensity of Expanded Point-Spread Function
for Five-Ring Spectral Window.

increase by the same factor as the central region. Because the magnitude of the sidelobes decreases monotonically with increasing distance from center, the magnitude of the periodic ambiguities decreases similarly with increasing radius. To a good approximation, beyond the first zero, the following approximation can be made (12):

$$J_0(4\pi r/\lambda) \cong \left(\frac{2\lambda}{\pi^2 r}\right)^{1/2} \cos(4\pi r/\lambda - \pi/4) \qquad (4\text{-}33)$$

The magnitude of the oscillations decreases as $(r)^{-1/2}$; the intensity of the periodic ambiguities therefore varies inversely with r. This behavior is verified in figures 4-28 and 4-29.

Spectral Sampling from Bistatic Measurements

In the preceding section, image reconstruction using data from discrete spectral rings was considered. Because the radius of a particular ring in the spatial frequency domain is $2/\lambda$, data for concentric rings are obtained from measurements at multiple temporal frequencies. In some applications, however, the available instrumentation will not permit changing the frequency of the transmitted signal. In this section, bistatic measurements are considered as a means of sampling the spatial spectrum over discrete concentric rings while retaining the convenience of using a single irradiating frequency.

In a bistatic geometry, the transmitter and receiver are spatially separated in contrast to a monostatic condition in which the two are collocated. The angle of separation, measured relative to the object center, is called the bistatic angle. In order to compare the two conditions, consider a point object with coordinates r, ϕ observed by monostatic and bistatic radars located at large equal distances R_0, and angles θ relative to the object center, as shown in figure 4-30. The elements of the bistatic radar are symmetrically displaced from the monostatic radar axis by angles $\beta/2$, where β is the bistatic angle which lies in the plane of the object. As the angle θ is varied, the phase of signals received by the radars will vary due to the changing round-trip distance. The complex envelope of signals received as a function of the angle for the monostatic and bistatic cases are, respectively:

$$G_M(\theta) = \exp(-j4\pi R_0\lambda^{-1})\exp\left\{ j2\pi\lambda^{-1}[2r\cos(\theta - \phi)] \right\} \qquad (4\text{-}34)$$

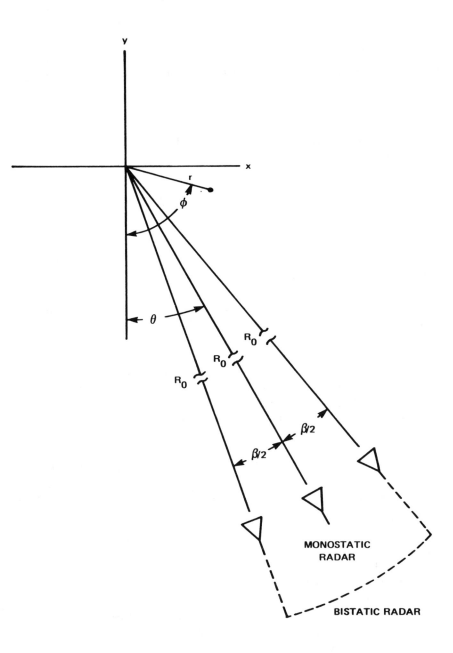

Figure 4-30. Monostatic and Bistatic Radar Configurations.

and

$$G_B(\theta) = \exp\left(-j4\pi R_0 \lambda^{-1}\right) \exp\left\{ j2\pi\lambda^{-1}[r\cos(\theta - \phi + \beta/2)\right.$$

$$\left. + r\cos(\theta - \phi - \beta/2)]\right\} \tag{4-35}$$

Applying the trigonometric identity $\cos(A+B) + \cos(A-B) = 2\cos A \cos B$ to equation (4-34) results in:

$$G_B(\theta) = \exp\left(-j4\pi R_0 \lambda^{-1}\right) \exp\left\{ j2\pi[(\cos\beta/2)\lambda^{-1} 2r\cos(\theta - \phi)]\right\}$$

$$\tag{4-36}$$

Comparison of equations (4-34) and (4-36) shows that the variable parts of the complex envelope have identical forms with the exception of the factor sec (β). This indicates that results of measurements with a bistatic angle β are equivalent to results that would be obtained with a monostatic geometry using a wavelength $\lambda\sec(\beta/2)$. This fact has been recognized in radar work and is known as the bistatic equivalence theorem. (13). The increase in effective wavelength resulting from the bistatic geometry allows measurements made with bistatic angles and a single temporal frequency to simulate measurements with multiple temporal frequencies. Two practical methods consist of: (1) sequential measurements made by displacing the transmitter and receiver about the monostatic axis, or (2) multiple measurements performed simultaneously using a single transmitter and multiple receivers. For this latter geometry, the effective aspect angle, θ, is defined by the bisector to the bistatic angle. The location of the transmitter or receiver is not restricted to the plane of the object. Other locations, including in a plane normal to that of the objects, are possible.

In order to test the practicality of the bistatic method, experimental data from a set of measurements performed on a two-point object using different bistatic angles were used to reconstruct an image. The test object was the pair of thin vertical rods described earlier in this chapter; the rods, separated by 34 cm, were irradiated by a wavelength of 6 cm from a distance of approximately 5 meters. Separate measurements were made using bistatic angles of 0, 58, 83, 102 and 120 degrees. This provided five spectral rings with radii uniformly spaced between $1/\lambda$ and $2/\lambda$, simulating the window function shown in figure 4-26. The images reconstructed from data obtained from measurements at each bistatic angle are shown in figure 4-31. The results illustrate the decreasing resolution as the bistatic angle is increased and the effective wavelength is increased accordingly. The combined image, formed by superimposing the complex amplitudes of the individual images, is shown in figure 4-32. The results show excellent correlation with the theoretical point-spread function of figure 4-26; the magnitudes and locations of the near side-lobes and of the ring artifacts are in close agreement. The image reconstructed from the five spectral rings shows a marked improvement over that from a single ring. The results demonstrate that the bistatic method is an effective way of simulating multi-frequency imaging while using a single irradiating frequency. A different reconstruction can be obtained by superimposing the magnitudes rather than the complex amplitude of the individual images. The result of this superposition, shown in figure 4-33, indicates a lesser improvement in the reconstruction than in the complex amplitude superposition. In this case, the sidelobe structure is averaged while in the complex amplitude superposition, some cancellation in the sidelobes occurs. We hasten to caution that the validity of the bistatic equivalence theorem requires that the amplitudes and location of the scattering centers not be altered by the bistatic geometry. This situation prevails for point objects and simple shapes; however, it is expected that when complex objects are observed at large bistatic angles, the bistatic equivalence may not be applicable, especially if corner reflectors are present or if the bistatic geometry alters the shadowing of some reflectors by others.

The results reported in this chapter are summarized as follows. When the complex envelope of signals reflected from a rotating object is Fourier-transformed to form a synthetic aperture, the point-spread function of the imaging process is degraded from the diffraction limit and is not space-invariant. Focusing the synthetic aperture leads to a space-invariant, diffraction-limited, point-spread function. The focusing operation is accomplished by transforming the complex envelope to a polar format which maps the amplitude of a point reflector as a planar wave in a two-dimensional transform space. An array of point reflectors, therefore, is mapped as a spectrum of planar wave components with amplitude and direction corresponding to the magnitude and location of each point. A subsequent two-dimensional Fourier transform decomposes the spectrum to reconstruct the spatial function. The focusing operation extends the required signal processing from a one-dimensional Fourier transform to a two-dimensional Fourier transform and therefore considerably increases the required computation. The focused aperture for a discrete frequency provides a high degree of spatial resolution but is

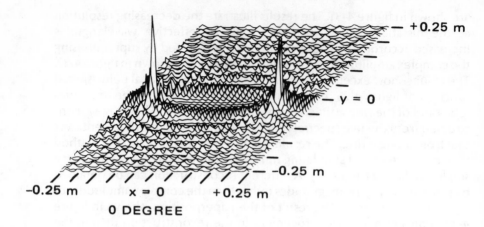

0 DEGREE

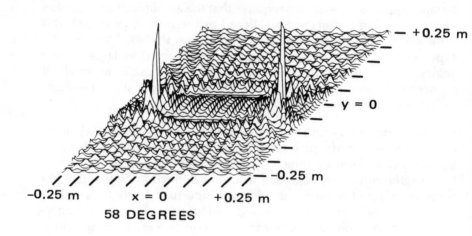

58 DEGREES

Figure 4-31. Intensity of Images of Two Point Objects
Reconstructed From Experimental Data for Bistatic
Angles of 0, 58, 83, 102, and 120 Degrees.

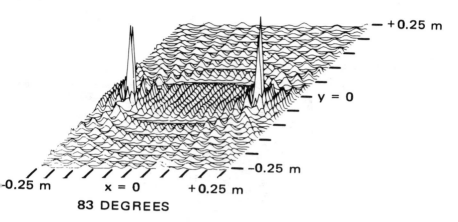

83 DEGREES

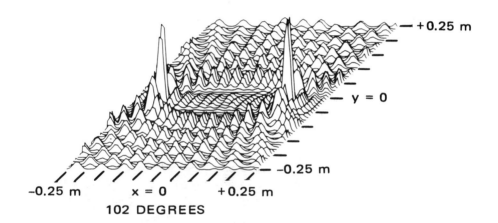

102 DEGREES

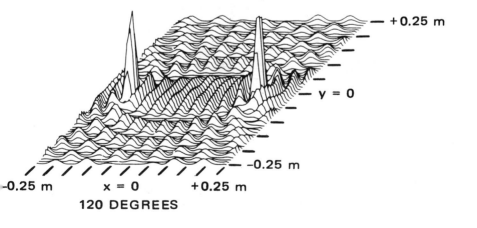

120 DEGREES

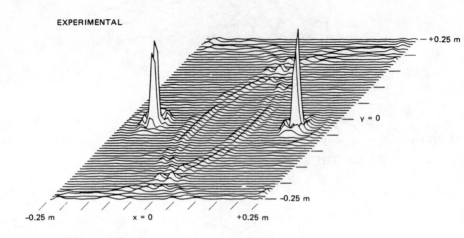

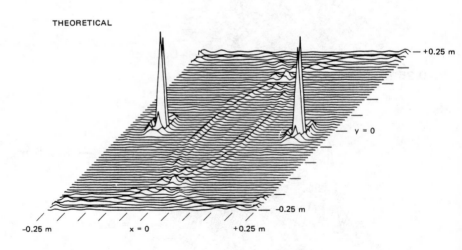

Figure 4-32. Intensity of Image Reconstructed From Complex Amplitude Superposition of Five Bistatic Images (λ=0.06 m).

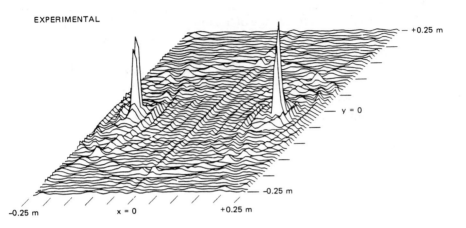

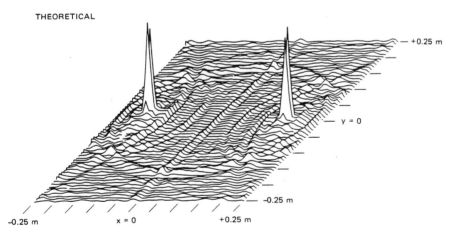

Figure 4-33. Intensity of Image Reconstructed From Magnitude
Superposition of Five Bistatic Images (λ=0.06 m).

usable for imaging only sparse arrays of point objects. This limitation is due to the presence of large sidelobes in the point-spread function which limits the dynamic range of the imaging process. The quality of the reconstruction is improved by using wide-band signals which provide a point-spread function with reduced sidelobes. The use of a number of discrete frequncies allows reduction of the sidelobes but induces low-level circular artifacts. Using discrete frequencies offers an alternative to wide-band signal instrumentation which is always more complex. In measurements restricted to fixed frequency CW operation, bistatic-angle diversity is a practical way of simulating multi-frequency measurements.

REFERENCES

1. Scudder, H. J. "Introduction to Computer Aided Tomography," Proc. IEEE, Vol. 66, No. 6, pp. 628-637, June 1978.
2. Kak, A. C. "Computerized Tomography with X-Ray, Emission, and Ultra-sound Sources," Proc. IEEE, Vol. 67, No. 9, pp. 1245-1271, September 1979.
3. Shepp, L. A., and B. F. Logan. "The Fourier Reconstruction of a Head Section," IEEE Trans. on Nuclear Science, Vol. NS-21, pp. 21-43, June 1974.
4. Cho, Z. H. "General View on 3-4 Image Reconstruction and Computerized Transverse Axial Tomography," IEEE Trans. on Nuclear Science, Vol. NS-21, pp. 44-71, June 1974.
5. Brown, W. M., and L. J. Porcello. "An Introduction to Synthetic Aperture Radar," IEEE Spectrum, p. 52, September 1969.
6. Chen, C. C., and H. C. Andrews. "Multi-Frequency Imaging of Radar Turntable Data," IEEE Trans. on Aerospace and Electronic Systems, Vol. AES-16, No. 1, pp. 15-22, January 1980.
7. Brown, W. M., and R. Fredericks. "Range-Doppler Imaging with Motion Through Resolution Cells," IEEE Trans. Aerospace and Electronic Systems, Vol. AES-5, pp. 98-102, January 1969.
8. Walker, J. L. "Range-Doppler Imaging of Rotating Objects," IEEE Trans. on Aerospace and Electronic Systems, Vol. AES-16, No. 1, pp. 23-52, January 1980.
9. Bracewell, R. *The Fourier Transform and Its Applications,* New York: McGraw-Hill, Co., 1965, pp. 244-250.
10. Harris, F. J. "On the Use of Windows for Harmonic Analyses with the Discrete Fourier Transform," Proc. IEEE, Vol. 66, No. 1, pp. 51-83, January 1978.
11. Simpson, R. G., H. H. Barrett, J. A. Suback, and H. D. Fisher. "Digital Processing of Annual Coded-Aperture Imagery," Optical Engineering, Vol. 14, No. 5, pp. 490-494, September-October 1975.
12. Ermert, H., and R. Karg. "Multi-Frequency Acoustical Holography," IEEE Trans. on Sonics and Ultrasonics, Vol. 26, No. 4, pp. 279-286, July 1979.
13. Kell, R. R. "On the Derivation of Bistatic RCS from Monostatic Measurements," Proc. IEEE, Vol. 53, pp. 983-988, August 1965.

An Iterative Method of
Image Reconstruction

The imaging methods presented in the preceding chapter were shown to provide exceptional resolution performance, demonstrated by point-spread functions with central regions of width less than one-half wavelength. The fidelity of the reconstructed images is limited by the high sidelobes of the point-spread functions which restrict the ability to distinguish images of objects with small magnitude in proximity to objects with large magnitude. The sidelobes, which result from spectral components missing in the measured spectral data, are increased when the measured spectrum is highly discontinuous. In some cases, the sidelobes can be suppressed by two-dimensional tapered windows. Windowing is an effective method for sidelobe control when many independent samples of data are available; the loss of measured data due to the windowing process is then not significant. When only small amounts of measured data are available, however, windowing can severely degrade the image.

The study of spectral estimation obtained from a small number of data samples has recently stimulated researchers to re-examine the concept of windowing and to consider alternatives for improved spectral estimates which are based on a more fundamental viewpoint. The method of maximum entropy appears to be the most active, judging by a number of significant papers which are presented in reference (1). The fundamental objection to windowing stems from the fact that, in the process, valid measured data are purposely altered and unmeasured data are arbitrarily set to zero. The philosophy of maximum entropy reconstruction, stated in (2), requires that "all extensions of measured data be consistent with the available data and a-priori information, and be maximally noncommittal regarding unavailable data." Assume that measurements are performed to obtain limited samples of a process and from these an attempt is made to describe, to the maximum extent possible, the process from which the samples were obtained. In order to estimate (reconstruct) the process we should: (1) accept the available data without alteration; (2) ensure that extrapolated data are consistent with all a priori information about the process; and (3) be as unbiased (noncommittal) as possible about data which are not measured and cannot be extrapolated. In spite of their apparent practicality, the above steps are often violated when data are processed to obtain spectral estimates. Consider, for example, the case illustrated in figure 4-24 of the preceding chapter in which spectral data, obtained over an annular region, are windowed to reduce sidelobes in the point-spread function. Two violations of the above premises are evident: (1) valid measured data corresponding to the edges of the annulus are altered; and (2) unmeasured data are arbitrarily set to zero. The latter is a clear violation in nearly all practical applications, in which the spatial spectrum being sampled corresponds to a spatially bounded project and is therefore a space-limited function.

A function g(x) is said to be space limited if it equals zero outside a finite region. The function g(x) is said to be band limited if its energy is finite and its Fourier transform $G(f_x)$ is zero outside a finite region (3), (4), (5). A band-limited function is analytic over the entire domain, meaning that the function has a derivative which is continuous in the domain (6). Because the Fourier transform relations are reciprocal, the spectrum of a space-limited function is analytic and the spatial function corresponding to a band-limited spectrum is analytic. Because an analytic function cannot be zero in a finite region, except for the trivial case when it is zero everywhere, a function cannot be simultaneously space limited and band limited. In the case of an object known to be spatially bounded, defining the spectrum to be zero everywhere outside a finite region poses a contradiction. With no additional information, the reconstruction should be based on a spectral function which coincides with the measured data over regions for which they are available and on an analytic extrapolation elsewhere.

An iterative procedure for extrapolating a given segment of the spectrum of an object known to be space limited is presented in (7), (8). Figure 5-1 is a representation of the computational procedure. The known portion of the spectrum is treated as band limited and Fourier transformed to reconstruct an image of the object. This is subsequently modified by setting the image to zero outside the known extent of the object. Subsequent to this modification, the object function is Fourier transformed and the resulting spectrum is corrected by replacing it with the known spectrum in the region where it is known. The process is repeated to successively extrapolate the measured spectrum and thus improve the reconstruction.

The above procedures are mathematically formulated for a one-dimensional case as follows: we seek to determine the Fourier transform $g(x)$ of a space-limited function $G(f_x)$ in terms of a known portion $G_0(f_x)$. The iteration starts with the transform of the known or measured spectrum:

$$g_0(x) = \int_F G_0(f_x) \exp\left[-j4\pi f_x x/\lambda\right]\, df_x \qquad (5\text{-}1)$$

where F is the region for which $G(f_x)$ is known.

The nth iteration proceeds by computing the inverse Fourier transform of:

$$f_n(x) = \begin{cases} g_{n-1}(x) & |x| \leqslant X \\ 0 & |x| > X \end{cases} \qquad (5\text{-}2)$$

which is given by:

$$F_n(f_x) = \int_X f_n(x) \exp\left[j4\pi f_x x/\lambda\right]\, d_x \qquad (5\text{-}3)$$

where X is the known spatial bound of the object.
Next, the following new function is formed:

$$G_n(f_x) = \begin{cases} G_0(f_x) & |f_x| \leqslant F \\ F_n(f_x) & |f_x| > F \end{cases} \qquad (5\text{-}4)$$

The nth step ends by computing the Fourier transform:

$$g_n(x) = \int_{-\infty}^{+\infty} G_n(f_x) \exp\left[-j4\pi f_x x/\lambda\right]\, df_x \qquad (5\text{-}5)$$

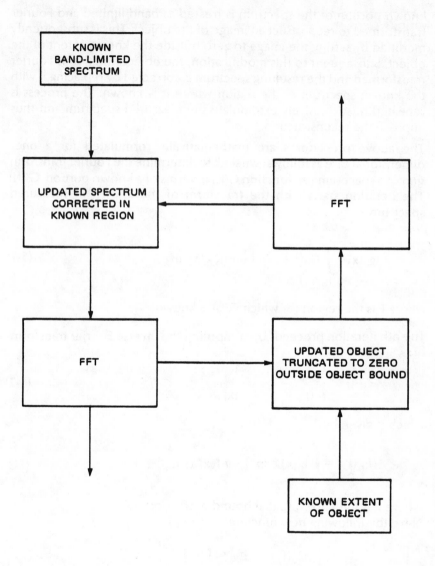

Figure 5-1. Computational Procedure of Iterative Reconstruction Method.

Figure 5-2 shows qualitative examples of spatial and spectral functions at various steps of the iteration. The double arrows indicate a Fourier transform operation. The limited spectrum can be viewed as the sum of the true spectrum and an error spectrum. The error spectrum is zero in the region where the true spectrum is known and is equal and opposite to the true spectrum outside this region. Because in the known spectral region the error spectrum is zero, it cannot be an analytic function and its Fourier transform is therefore continuous. When the spectrum is transformed to the object domain, the algorithm sets the error object to zero outside the known extent of the subject. The truncation, however, has no effect on the true object because its extent is equal to or less than the truncation window.

Because the energies of the error spectrum and error object are identical prior to the truncation (by Parseval's theorem), the error energy must be reduced by the truncation process. The subsequent Fourier transform of the truncated error object is analytic and the new error spectrum therefore contains energy in the region where the spectrum is known. Correcting the spectrum in the known region reduces the error energy in this region to zero and creates an error spectrum which is band-limited and therefore has an analytic transform. Thus, at each correction, first by the truncation in the object domain and then by the substitution of the known spectrum, the error energy is reduced. The fidelity of the reconstructed image depends on the error-reduction features of the algorithm. Reduction of the error energy outside the object bound follows directly from the spatial truncation and occurs at each iteration. Reduction of the error energy inside the object bound, however, is less direct and depends on the nature of the error spectrum. As the error spectrum becomes less discontinuous, more of the energy content of its transform, the error object, falls inside the object bound and is unaffected by the subsequent truncation. As this condition is approached, the error reduction process terminates. The reduction of the error energy, and therefore the fidelity of the reconstruction, are greater if the extent of the true object is used than if the object size is overestimated. The preceding formulation was one-dimensional; however, the identical algorithm strategy can be applied to multi-dimensional transforms. The error-reducing feature of the algorithm is maintained because the Fourier transform is a linear operator, and we therefore expect the algorithm to provide analytic extrapolations of two-dimensional, band-limited spectra.

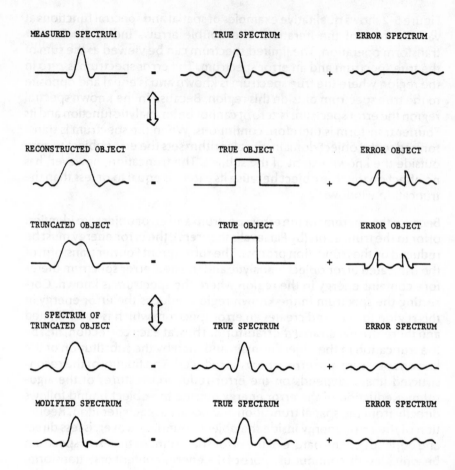

Figure 5-2. Qualitative Examples of Spatial and Spectral Functions at Various Steps of the Iteration.

In order to test the above assertions, the algorithm was used to reconstruct the image of a disc-shaped object from a limited section of the spatial spectrum. The data input to the algorithm were computed from an idealized, noise-free spectrum shown in figure 5-3. The total object space is a square area with side 16λ; the spectrum is plotted on a square field with side $8/\lambda$. The amplitude scale for both functions is logarithmic over a range of 50 db. The object is a circular disc of radius λ. Figures 5-4 through 5-6 show results at intermediate steps of the algorithm for the first, tenth, and twentieth iteration. The scales for these and all subsequent plots are identical to those of figure 5-3. Each plot is a square array of 128 cells on the side. The Fourier transform operations used in the algorithm are two-dimensional FFT's operating on arrays of 128 by 128 complex point pairs. In this example, as in (7), (8), the dimension of the object was assumed to be precisely known and the object bound made to coincide with the extent of the object. The results show that the algorithm converges rapidly to a good reconstruction of the postulated object, evidenced by the faithful reconstruction inside the object bound and the low level of the artifacts outside the object bound. The identical reconstruction was repeated with the size of the object overestimated by a factor of two; that is, the true object was represented by a disk of radius λ while the object bound was a circular region of radius 2λ. Figures 5-7 through 5-9 show intermediate steps of the algorithm after the first, tenth, and twentieth interations, respectively. These results show that the reconstruction does not converge to the true object; after twenty iterations, the reconstruction inside the object bound is nearly identical to that obtained in the first iteration. The error energy outside the object bound is reduced drastically while the error energy in the region between the true object and the object bound is essentially unaltered. This illustrates the situation described earlier in which the error spectrum extends smoothly over a large region of the spectrum, and its transform, the error object, falls essentially within the object bound. This condition precludes a reduction in the subsequent truncation. The preceding results demonstrate that the algorithm is equally effective in two-dimensional and in one-dimensional applications. The Fourier transform operations in the examples were performed using two-dimensional FFT subroutines. However, the results are similar to those for a continuous formulation. We therefore expect the algorithm to be effective in two-dimensional, sampled data applications.

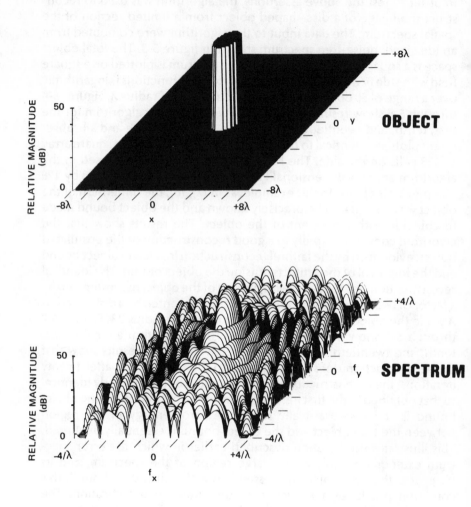

Figure 5-3. True Object and Spectrum of Idealized Case Used to Test the Reconstruction Algorithm.

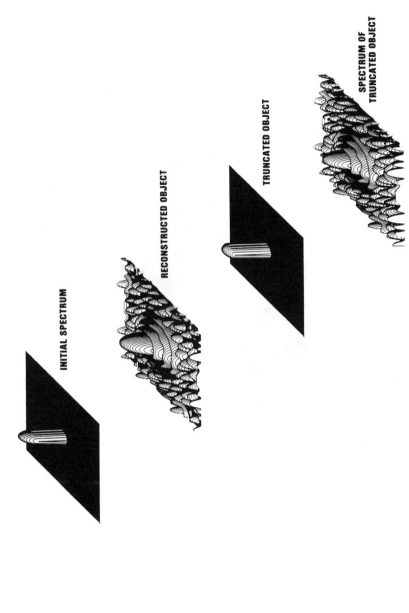

Figure 5-4. Spatial and Spectral Responses for First Iteration of the Algorithm Applied to a Partial Spectrum of a Disc Object (Radius of Object Truncation = λ).

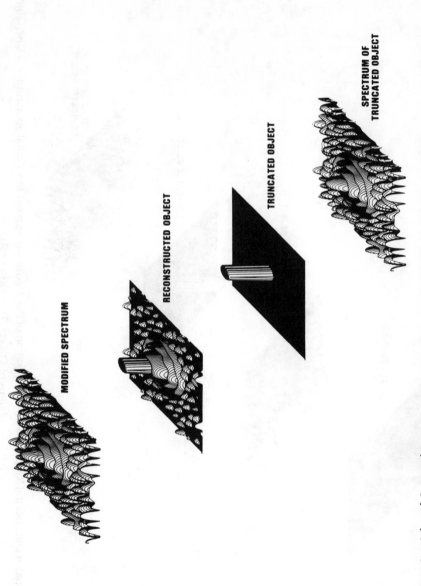

Figure 5-5. Spatial and Spectral Responses for Tenth Iteration of the Algorithm Applied to a Partial Spectrum of a Disc Object (Radius of Object Truncation = λ).

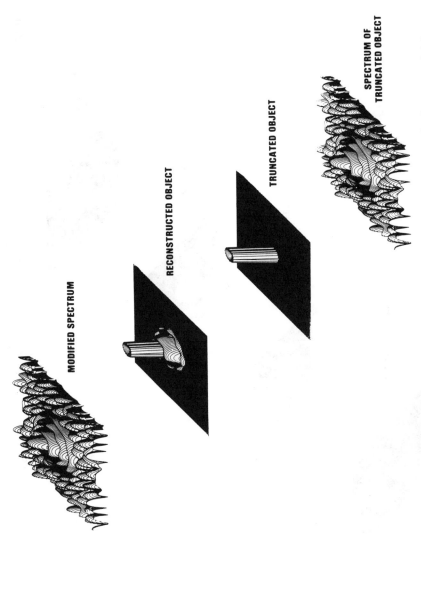

Figure 5-6. Spatial and Spectral Responses for Twentieth Iteration of the Algorithm Applied to a Partial Spectrum of a Disc Object (Radius of Object Truncation = λ).

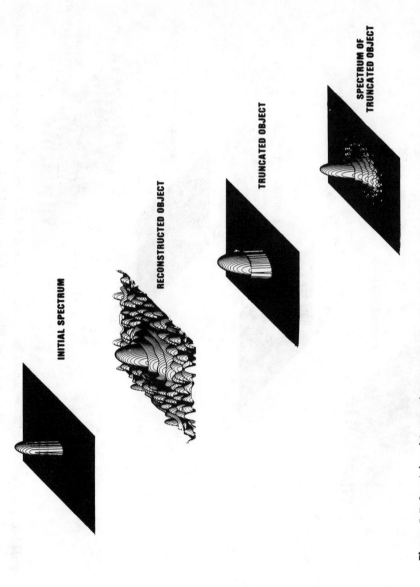

Figure 5-7. Spatial and Spectral Responses for First Iteration of the Algorithm Applied to a Partial Spectrum of a Disc Object (Radius of Object Truncation = 2λ).

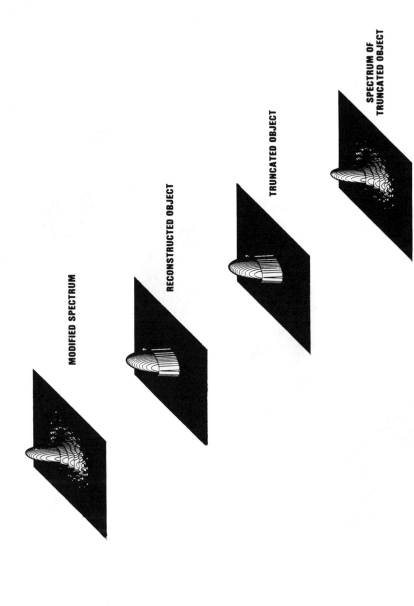

Figure 5-8. Spatial and Spectral Responses for Tenth Iteration of the Algorithm Applied to a Partial Spectrum of a Disc Object (Radius of Object Truncation = 2λ).

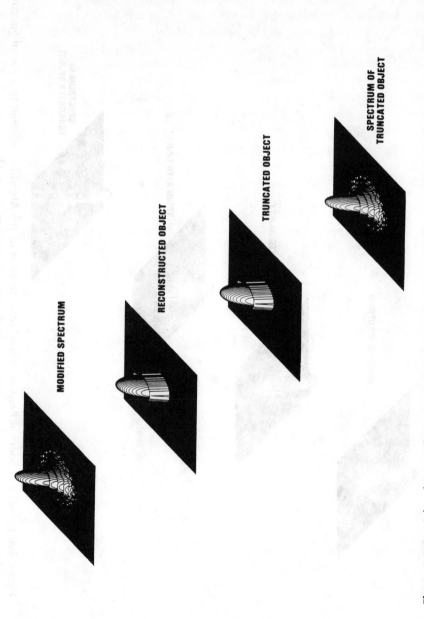

Figure 5-9. Spatial and Spectral Responses for Twentieth Iteration of the Algorithm Applied to a Partial Spectrum of a Disc Object (Radius of Object Truncation 1.3)).

In the following section, we consider using the iterative algorithm just described to improve the images reconstructed from the focused synthetic aperture process described in chapter IV. It was shown there that CW reflection measurements, properly transformed, yielded a sample of the spatial spectrum over a circular ring of radius 2λ and that the reconstruction from such data exhibited a radially symmetric point-spread function $J_0(4\pi r/\lambda)$. Postulating the spectrum to have zero value everywhere except on the circular ring imposed a band limit and destroyed the analyticity of the spectrum. As a consequence, the resulting spatial function was analytic with extended sidelobe artifacts.

The extension of a truncated spectrum beyond its given limits is the basis for achieving resolution beyond the diffraction limit. Such concepts are theoretically sound, as documented in (9)-(11); however, "super-resolution" has not been highly successful in practical cases due mainly to limitations of noise and measurement precision. In this application, the objective is not to construct perfectly the spectrum out to some new diffraction limit, but only to extend it sufficiently to improve the reconstructed image by reducing the sidelobe artifacts. As shown in chapter IV, reduction of the sidelobes in the point-spread function results from broadening the spectral ring; such improvements therefore appear more feasible than the elusive goal of obtaining "super-resolution."

In order to test the effectiveness of the algorithm, simulated data representing ideal measurements of reflections from a point located at the center of the object space were used. This is representative of practical microwave images of complex objects which behave as arrays of isolated point-like reflectors. In such cases, the only available a priori information on the object bound consists of the overall object dimensions. While we can reliably establish the bounds of the array, the bounds of the individual reflectors will necessarily be overestimated because their location inside the array boundary cannot be established a priori. Unit-amplitude data were arrayed on a circular ring in a 128 x 128-point rectangular grid spanning the frequency domain. The input data array was formed by placing ones in each cell for which the radial coordinate was between 31.5 and 32.5 cell dimensions, and zeros elsewhere. The data input to the algorithm represent a spectral ring of radius $2/\lambda$ arrayed at the center of a square spectral field of dimension $8/\lambda$; the corresponding extent of the spatial domain field is 16λ.

Figures 5-10 through 5-12 show intermediate steps in the reconstruction for the first, tenth, and twentieth iterations. The object bound is a circular region of radius λ. At the end of the first iteration, the spectrum

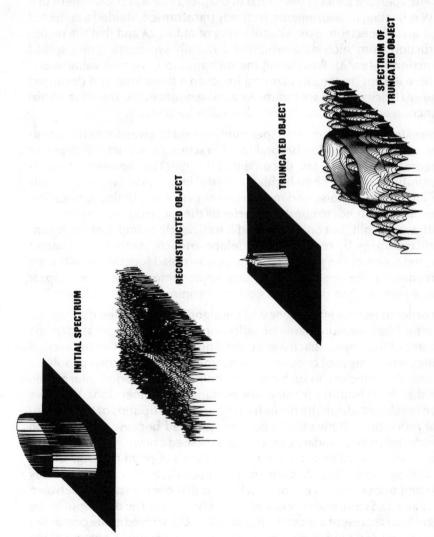

Figure 5-10. Spatial and Spectral Responses for First Iteration of the

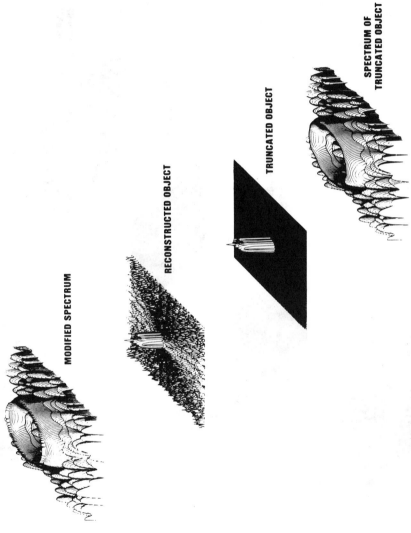

Figure 5-11. Spatial and Spectral Responses for Tenth Iteration of the Algorithm Applied to a Ring Spectrum (Radius of Object Truncation = λ).

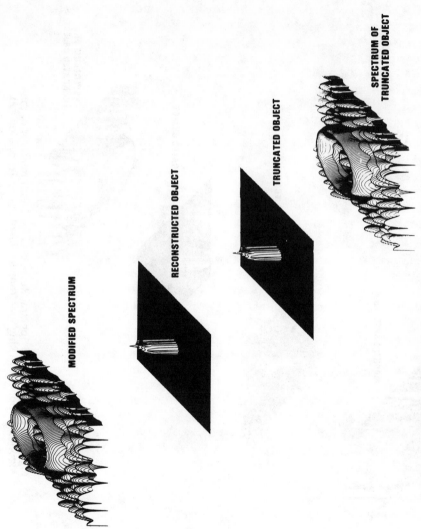

Figure 5-12. Spatial and Spectral Responses for Twentieth Iteration of the Algorithm Applied to a Ring Spectrum (Radius of Object Truncation = a)

is broadened to a width which is inversely proportional to the object bound. After the tenth iteration, the response outside the object bound is diminished considerably but the spectrum is not materially broadened. After twenty iterations, the response outside the object bound is decreased further but the spectrum is not altered. The reconstruction image inside the object bound is essentially unaltered by the algorithm. Figures 5-13 through 5-15 show similar results for an object bound at radius 4λ. The reconstruction outside the object bound is improved significantly; inside the object bound, however, no significant improvement is noted.

Figures 5-16 through 5-18 show reconstruction from an annular spectral region with inner and outer radii of $1.5/\lambda$ and $2/\lambda$, respectively. The object bound is a circular region of radius 4λ. The results are very similar to the preceding case of the ring spectrum: significant improvement in the reconstruction outside the object bound but no significant improvement inside the object bound.

The results obtained from applying the algorithm to simulated data demonstrate a rapid convergence to an accurate reconstruction when the postulated object extent coincides with the size and shape of the true object. When the object size is overestimated, the reconstruction inside the object bound is not significantly improved by successive iterations. In the limiting case of reconstructing the image of a point object contained in a large object bound, the improvement inside the object bound is insignificant. The reconstruction outside the object bound rapidly approaches zero; however, this can be achieved directly from the bounded-object condition without recourse to the algorithm. The possibility of achieving significant improvements after many iterations was eliminated by continuing the reconstruction of the ring spectrum to 100 iterations without significant effects. As a result of these findings, the algorithm is considered useful only in restricted cases where the object bound is accurately known a priori. The algorithm does not improve the quality of images reconstructed from focused synthetic aperture measurements of complex objects consisting of arrays of point reflectors. The search for alternative methods of extrapolating space-limited spectra in such applications should be the subject of continued research.

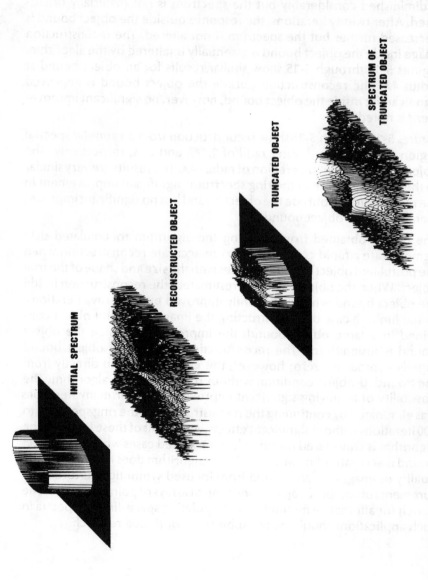

Figure 5-13. Spatial and Spectral Responses for First Iteration of the Algorithm Applied to a Ring Spectrum

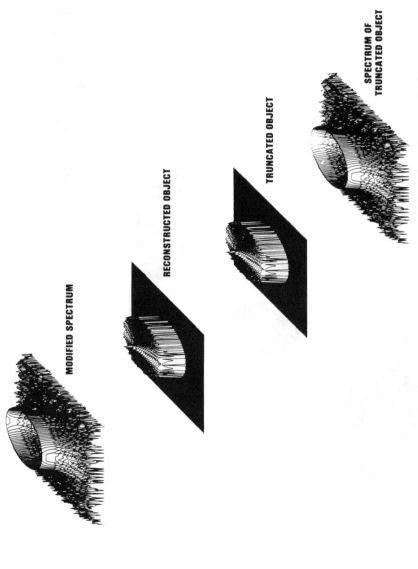

Figure 5-14. Spatial and Spectral Responses for Tenth Iteration of the Algorithm Applied to a Ring Spectrum (Radius of Object Truncation = 4λ).

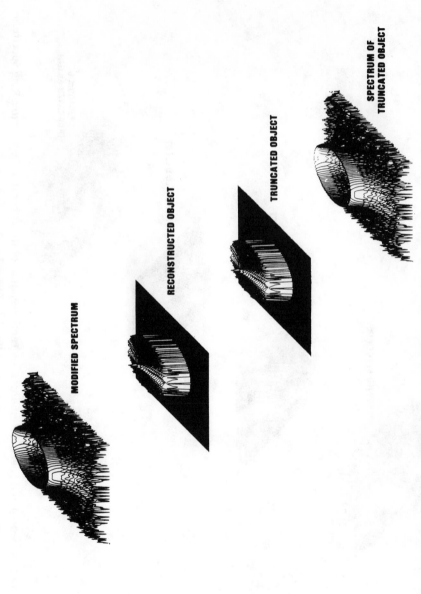

Figure 5-15. Spatial and Spectral Responses for Twentieth Iteration of the Algorithm Applied to a Ring Spectrum

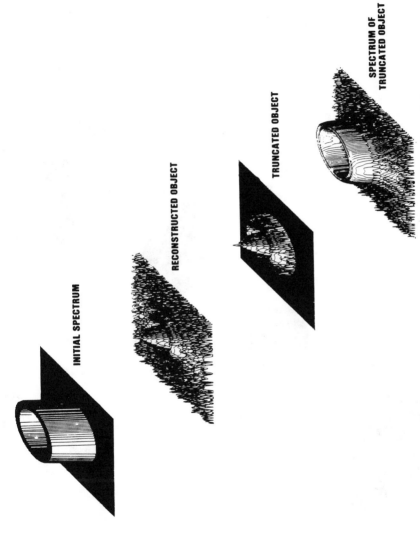

Figure 5-16. Spatial and Spectral Responses for First Iteration of the Algorithm Applied to a Annular Spectrum (Radius of Object = 4λ).

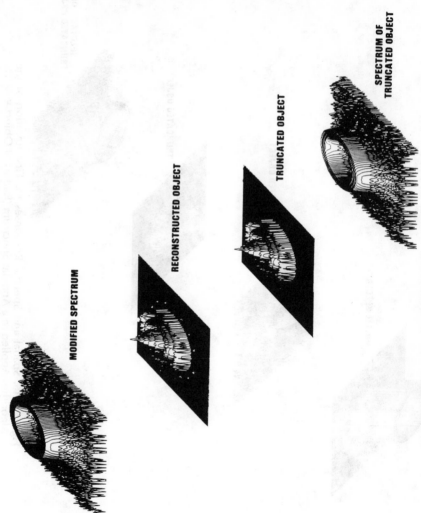

Figure 5-17. Spatial and Spectral Responses for Tenth Iteration of the Algorithm Applied to a Annular Spectrum (Radius of Object = 4λ)

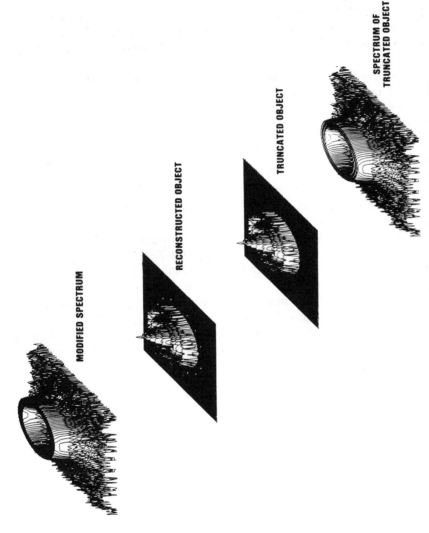

Figure 5-18. Spatial and Spectral Responses for Twentieth Iteration of the Algorithm Applied to a Annular Spectrum (Radius of Object = 4λ).

REFERENCES

1. Childers, D. G., Ed. *Modern Spectrum Analysis*. New York: IEEE Press, 1978.
2. Ables, J. G. "Maximum Entropy Spectral Analysis, " Astronomy and Astrophysics Supplement Series, Vol. 15, pp. 383-393, June 1974.
3. Papoulis, A. *Signal Analysis*. New York: McGraw-Hill Co., 1977, pp. 183-191.
4. Bracewell, R. *The Fourier Transform and Its Applications*. New York: McGraw-Hill Co., 1965, pp. 160-163.
5. Temes, G. C., V. Barcilon, and F. C. Marshall. "The Optimization of Bandlimited Systems," Proc. IEEE, Vol. 61, No. 2, pp. 196-205, February 1973.
6. Kaplan, W. *Introduction to Analytic Functions*. Reading: Addison Wesley Co., 1966, pp. 35-36.
7. Papoulis, A. "A New Algorithm in Spectral Analysis and Band-Limited Extrapolation," IEEE Trans. on Circuits and Systems, Vol. CAS-22, No. 9, pp. 735-742, September 1975.
8. Gerchberg, R. W. "Super-Resolution Through Error Energy Reduction," Optica Acta, Vol. 21, No. 9, pp. 709-720, 1974.
9. di Francia, G. T. "Resolving Power and Information," Journal of the Optical Society of America, Vol. 45, No. 7, pp. 497-501, July 1955.
10. Harris, J. L. "Diffraction and Resolving Power," Journal of the Optical Society of America, Vol. 54, No. 7, pp. 931-936, July 1964.
11. Barnes, C. W. "Object Restoration in a Diffraction-Limited Imaging System, " Journal of the Optical Society of America, Vol. 56, No. 5, pp. 575-578, May 1966.

Appendix

Application of Tapered Windows to Annular-Shaped Spectra

As shown in chapter IV, processing a two-dimensional record of signals reflected from a rotating object as a function of frequency and rotation angle provides a two-dimensional spatial image of the object's reflectivity density. When mapped in a polar format, the recorded data represent a spatial spectrum related to the object image by a two-dimensional Fourier transform. Because in practical measurements the range of irradiating frequencies is restricted to a limited bandwidth, the available data represent a sample of the spatial spectrum over an annular region. The point-spread function of the imaging process, given by the Fourier transform of the annular spectrum, consists of a central peak surrounded by circular sidelobes which generally increase as the width of the spectral annulus decreases. In view of practical limitations, we therefore seek the best image reconstruction from narrow annular spectra. In an analogous situation, spectral estimates obtained from time-limited signals are improved by the use of tapered windows which have been the subject of considerable research (1). The mechanism for reducing the spectral sidelobes is intuitively acceptable: discontinuities in the time function represent high frequencies in the spectrum and tapering to reduce discontinuities therefore reduces high frequency sidelobes. As a result, smoothly tapered time windows reduce spectral sidelobes.

Indiscriminate applications of tapered windows to two-dimensional functions may not produce the desired reduction in the sidelobes of the transforms, as shown in the example of figure 4-24. In this case, a cosinusoidal taper applied to an annular spectrum increased the sidelobe level rather than decreasing it. On first consideration, this appears contrary to the intuitive notion that smooth tapers in one domain decrease sidelobes in the other. This apparent paradox is resolved in the analysis which follows.

Because the spectra considered are radially symmetric functions, the corresponding spatial functions, being related by a Fourier transform, are also radially symmetric (2). This fact allows each of the functions to be characterized by a single variable expressing the variations as a function of radial distance. Although the functions are expressible by a single variable, they are nevertheless two-dimensional functions and must be treated as such when transformed. The Fourier transform can be reduced to a one-dimensional integral by use of the projection theorem which has been widely applied in the analysis of tomographic imaging (3) - (5). The theorem states that a central slice of the two-dimensional spectrum at an angle θ can be obtained by a one-dimensional transform of the projection of the spatial function onto an axis oriented at the same angle θ. The theorem can be proved by the following arguments.

Consider a two-dimensional function f(x,y) expressed in terms of coordinates u, v rotated about x, y by the angle θ as shown in figure 1.

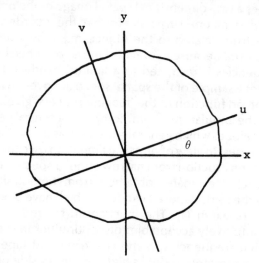

Figure 1. Rotated Coordinate System.

The function $f(x,y)$ can be expressed in terms of u and v by the rotational transformation $u = x \cos\theta + \sin\theta$, $v = -x \sin\theta + y \cos\theta$.

$f(x,y) = f(u \cos\theta - v \sin\theta, u \sin\theta + v \cos\theta)$

$$f(x,y) = f(u \cos\theta - v \sin\theta, u \sin\theta + v \cos\theta) \tag{1}$$

The projection of $f(x,y)$ onto the u axis corresponding to the angle θ is a one-dimensional function of the single variable u.

$$p(u;\theta) = \int f(u \cos\theta - v \sin\theta, u \sin\theta + v \cos\theta)\, dv \tag{2}$$

The notation $p(u;\theta)$ indicates that the projection is a function of the variable u and is unique for each θ. The one-dimensional Fourier transform of $p(u;\theta)$ with respect to the variable u is:

$$P(\omega;\theta) = \int p(u;\theta) e^{-j\omega u}\, du \tag{3}$$

The Fourier transform of the function $f(x,y)$, expressed in polar coordinates, is:

$$F(\omega,\theta) = \iint f(x,y) e^{-j(\omega x \cos\theta + \omega y \sin\theta)}\, dx\, dy \tag{4}$$

where ω and θ are the radial and angular coordinates in the spatial frequency plane. Equation (4), expressed in terms of the coordinate rotation relations, can be rewritten as:

$$F(\omega,\theta) = \iint f(u \cos\theta - v \sin\theta, u \sin\theta + v \cos\theta) e^{-ju\omega}\, du\, dv \tag{5}$$

$$F(\omega,\theta) = \int p(u;\theta) e^{-ju\omega}\, du \tag{6}$$

Equation (6) is thus shown to be equal to equation (3) yielding:

$$F(\omega,\theta) = P(\omega;\theta) \tag{7}$$

Equation (7), termed the projection theorem, states that the one-dimensional Fourier transform of the projection with orientation θ is a section of the two-dimensional Fourier transform of $f(x,y)$ passing through the origin and subtending an angle θ with the ω_x axis.

In our application, we use the above theorem by computing the projection of the annular spectrum and then performing a one-dimensional Fourier transform on the result. Because both spatial and spectral functions are radially symmetric, the result completely characterizes the point-spread function. The recognition that the radial behavior of the spatial function is determined by the Fourier transform of the projection of the spectral function is significant. All the features of one-dimensional tapered windows which link smooth tapers to sidelobe reduction are pertinent to the two-dimensional case if they are applied to the one-dimensional projection of the spectral function. Thus, smoothing the projection of the annular spectrum determines the sidelobe structure of the point-spread function in the radial direction.

The upper plot of figure 2 shows the cross section of an annular spectrum with unit amplitude for radial values between 32 and 64 units and zero elsewhere. The projection of the annular spectrum is shown on the central plot and the Fourier transform of the projection on the lower plot. The Fourier transform, representing the point-spread function corresponding to the annular spectrum, has sidelobes of magnitude 0.3 relative to the peak. The two central peaks of the projection are the cause of the relatively high side-lobes in the point-spread function. As the radial extent of the annulus decreases, the peaks become more pronounced and the side-lobes in the point-spread function increase. Figures 3-5 show effects of radially tapering the annulus with functions of the form $\cos^{\alpha}(\cdot)$ for α of 1, 2, and 0.5. Figures 6-8 show the effects of similar tapers applied to the inner and outer edges of the annulus. Each of the tapers results in an increase in the level of the side-lobe adjacent to the peak.

Note that in each case, the tapering enhances the peaked behavior of the projection thereby raising the sidelobe level of the corresponding spectrum. A number of additional tapers were tested for which results are not shown. These included functions which provided smooth tapers radially outward and inward. For each taper tested, the sidelobes of the point-spread function were increased over those for the uniform annular spectrum. Although the results are not analytically conclusive, they support the assertion that smooth tapering of the annular spectra does not decrease the sidelobes of the point-spread function but rather increases them.

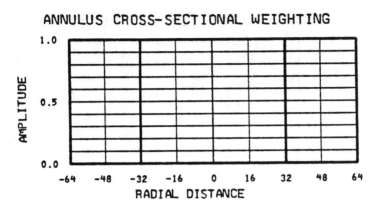

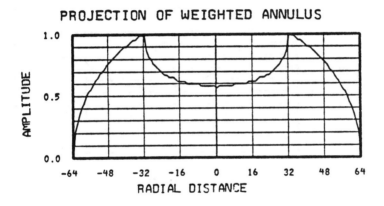

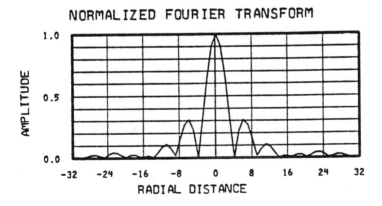

Figure 2. Annular Spectrum With Uniform Weighting.

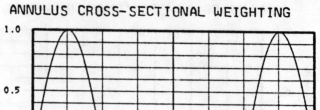

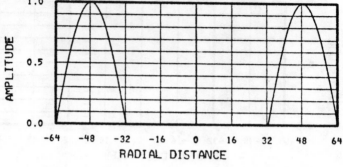

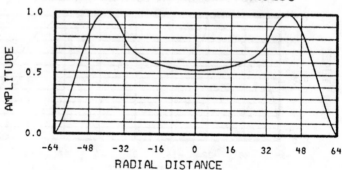

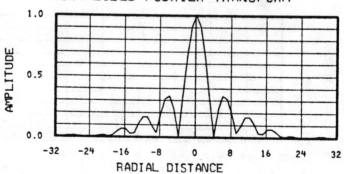

Figure 3. Annular Spectrum With Cos (·) Tapered Weighting.

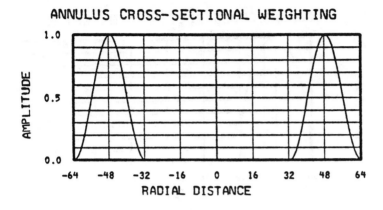

ANNULUS CROSS-SECTIONAL WEIGHTING

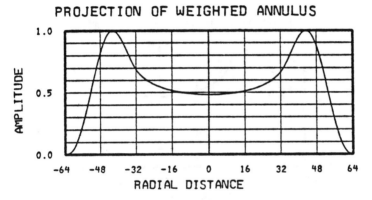

PROJECTION OF WEIGHTED ANNULUS

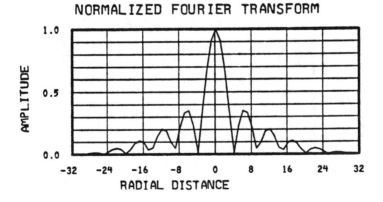

NORMALIZED FOURIER TRANSFORM

Figure 4. Annular Spectrum With Cos^2 (·) Tapered Weighting.

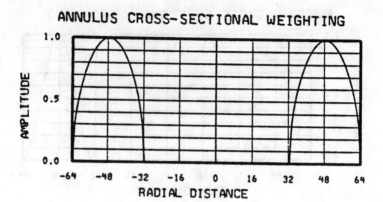

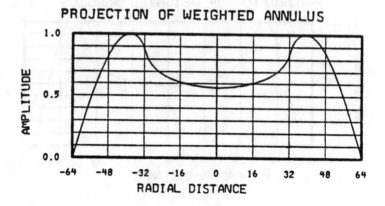

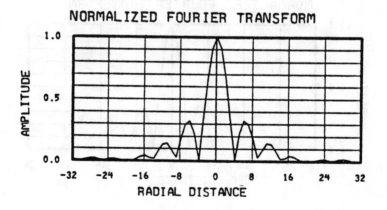

Figure 5. Annular Spectrum With Cos$^{1/2}$ (·) Tapered Weighting.

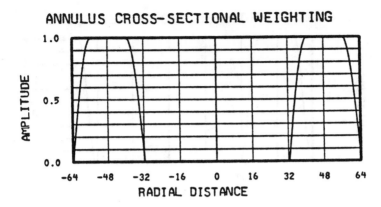

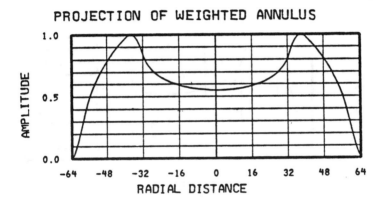

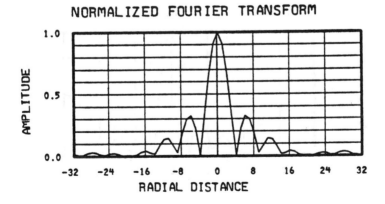

Figure 6. Annular Spectrum With Cos (·) Tapered Edge Weighting.

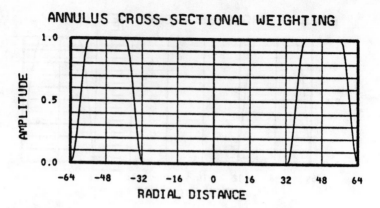

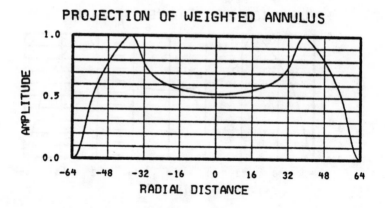

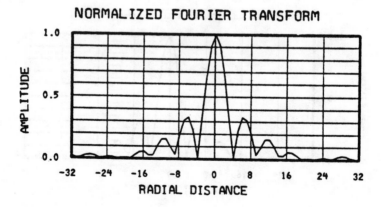

Figure 7. Annular Spectrum With Cos² (·) Tapered Edge Weighting.

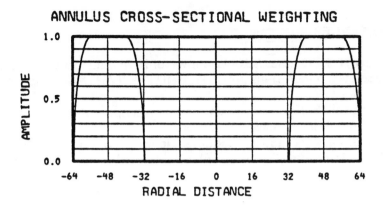

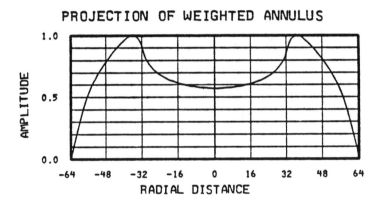

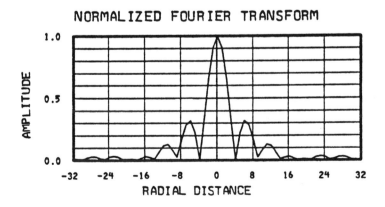

Figure 8. Annular Spectrum With Cos$^{1/2}$ (·) Tapered Edge Weighting.

REFERENCES

1. Harris, F. J. "On the Use of Windows for Harmonic Analyses with the Discrete Fourier Transform," Proc. IEEE, Vol. 66, No. 1, pp. 51-83, January 1978.
2. Bracewell, R. *The Fourier Transform and Its Applications.* New York: McGraw-Hill Co., 1965, pp. 244-250.
3. Bracewell, R. N., and A. C. Riddle. "Inversion of Fan-Beam Scans in Radio Astronomy," The Astrophysical Journal, Vol. 150, pp. 427-434, November 1967.
4. Shepp, L.A., and J.B. Kruskal. "Computerized Tomography: The New Medical X-Ray Technology," American Mathematical Monthly, Vol. 85, No. 6, pp. 420-439, June 1978.
5. Mersereau, R. M., and A. V. Oppenheim. "Digital Reconstruction of Multidimensional Signals from their Projections," Proc. IEEE, Vol. 62, No. 10, pp. 1319-1338, October 1974.

Index